The Quantum Vacuum

The Quantum Vacuum

*A Scientific and Philosophical Concept,
from Electrodynamics to String Theory and the
Geometry of the Microscopic World*

LUCIANO BOI

The Johns Hopkins University Press
Baltimore

© 2011 The Johns Hopkins University Press
All rights reserved. Published 2011
Printed in the United States of America on acid-free paper
2 4 6 8 9 7 5 3 1

The Johns Hopkins University Press
2715 North Charles Street
Baltimore, Maryland 21218-4363
www.press.jhu.edu

Library of Congress Cataloging-in-Publication Data

Boi, L. (Luciano), 1957–
The quantum vacuum : a scientific and philosophical concept, from
electrodynamics to string theory and the geometry of the microscopic
world / Luciano Boi.
p. cm.
Includes bibliographical references and index.
ISBN-13: 978-1-4214-0247-5 (hardcover : alk. paper)
ISBN-10: 1-4214-0247-5 (hardcover : alk. paper)
1. Vacuum.   2. Quantum field theory.   I. Title.
QC174.52.V33B65 2011
540.14'3–dc22   2011000991

A catalog record for this book is available from the British Library.

*Special discounts are available for bulk purchases of this book. For more
information, please contact Special Sales at 410-516-6936 or
specialsales@press.jhu.edu.*

The Johns Hopkins University Press uses environmentally friendly
book materials, including recycled text paper that is composed of at
least 30 percent post-consumer waste, whenever possible.

Nature has no aversion to the vacuum.

B. Pascal, *Expériences nouvelles touchant le vide*, 1647

At the very smallest scale of distances, the proper starting point for
the description of physics would therefore seem to be, not the elementary
particle, but the vacuum, complex in geometry and rich in dynamics.

J. A. Wheeler, *Geometrodynamics*, 1963

The quantum vacuum does indeed have interesting geometrical features,
but these relate not to the traditional geometry of Euclid, Riemann, etc.,
involving measurement, but to that modern branch of the subject known
as topology, which is concerned with qualitative properties of space.
Unlike measurement, which can be conducted on a small scale, topological
features are visible only on a "global" scale.

M. Atiyah, *Topology of the Vacuum*, 1991

Contents

Acknowledgments

For useful comments and stimulating conversations on many of the topics addressed in the present work, I would like to thank Sergio Albeverio, Sisir Roy, Ernesto Carafoli, Giuseppe O. Longo, Antonio Saggion, Gabriele Veneziano, Ugo Bruzzo, Salvatore Califano, and Arthur Miller.

Acknowledgment

Prologue

Answers to some of the long-standing issues in twentieth-century theoretical physics, such as the cosmic singularity, the black matter, and energy, appear to be closely related to the problem of the quantum vacuum and its fluctuations. The properties of the universe arise from "nothing," but this nothing is the quantum vacuum, which is a very different kind of nothing. If we examine a piece of "empty" space, we see that it is not truly empty. Rather, it is filled with space-time networks; for example, spacetime possesses immaterial curvature and structure, and obeys the laws of quantum physics. Thus, it is filled with potential particles, unit pairs of virtual matter and anti-matter, as well as potential properties at the quantum level. The quantum vacuum is in fact the ground state of energy for the universe, the lowest possible level of existence. To a first approximation, this is simply a state containing no particles, but even an ideal vacuum, thought of as the complete absence of anything, will not in practice remain empty.

One reason that perfect vacuum is impossible lies in the Heisenberg uncertainty principle, which states that no particle can ever have an exact position. Each atom exists as a probability function of space, and space has a certain nonzero value everywhere in a given volume. More fundamentally, quantum mechanics predicts that vacuum energy will differ quite substantially from its naïve classical value. The quantum correction to energy, called zero-point energy, consists of energies of virtual particles each having a brief existence. This is called *vacuum fluctuation*. In other words, quantum fluctuation is the temporary emergence of energetic particles from nothing, as allowed by the uncertainty principle. Vacuum fluctuations may also be related to the so-called cosmological constant of cosmology. (The best evidence for vacuum fluctuations is the Casimir effect and the Lamb shift.)

The quantum vacuum is thus, curiously, a more complex entity than the usual macroscopic entities, and it is far from being featureless and far from being empty; it is in fact the source of all potentiality. For example, quantum entities have both wave and particle characteristics. It is from the quantum vacuum that such characteristics emerge. Such particles "stand out" from the vacuum, and leave their signature on objects in the real universe. In this sense, the universe is not filled by the quantum vacuum. Rather, *it is written on it*, the immaterial substratum of all physical existence. I will argue that, in some sense, with respect to the origin of the universe, the quantum vacuum must have been the source of the laws of nature and the properties that we observe today.

Chapter 1

Introduction

The Vacuum *as a Scientific and Philosophical Concept*

A vacuum is a volume of space that is essentially empty of matter, such that its gaseous pressure is far, far less than standard atmospheric pressure. The root of the word *vacuum* is the Latin adjective *vacuus*, which means "empty," but space can never be perfectly empty. A perfect vacuum with a gaseous pressure of absolute zero is an idealized concept never observed in practice, not least because quantum theory predicts that no volume of space can be perfectly empty in this way.

The terms "nothing," "void," and "vacuum" usually suggest uninteresting empty space, but to modern quantum physicists the vacuum has turned out to be rich with complex and unexpected behavior. They envisage it as a state of minimum energy where quantum fluctuations, consistent with the uncertainty principle of Werner Heisenberg, can lead to the temporary formation of particle-antiparticle pairs. As will be explained in some detail later, according to quantum electrodynamics (QED) the vacuum is richly provided with electron-positron fields. Actual electron-positron pairs are created when energetic photons, represented by the electromagnetic field, interact with these fields. Virtual electron-positron pairs, however, can also exist for unimaginably minute durations (less than a Planck time $t_p = \sqrt{G\hbar / c^5} = 5.39124 \times 10^{-44}$ sec), as dictated by the uncertainty principle.[1]

Since ancient Greek times, vacuum has been a frequent topic of philosophical debate, but it was not studied empirically until the seventeenth century, when the theories of atmospherical pressure of Evangelista Torricelli, a contemporary of Galileo, would lead to the development of experimental techniques. Recall that Greek philosophers did not like to admit the existence of a vacuum, asking themselves "How can 'nothing' be something?" Plato found the idea of a vacuum inconceivable. He believed that all physical things were instantiations of an abstract ideal, and he could not conceive of an "ideal" form of a vacuum. Similarly, Aristotle considered the creation of a vacuum impossible—nothing could not be something. Later Greek philosophers thought that a vacuum could exist outside the cosmos, but not within it.

We may note here that in Aristotelian physics, the medium (such as water or air, etc.) plays a dual role: (1) providing a cause for motion (toward a proper or natural place) and (2) resisting the motion (let it be infinite). In fig. 1 we have a graphic representation of Aristotle's argument. Body *A* moves (in natural motion) two equal distances, once through water (*B*) in a longer time *C* and once through air (*D*) in a shorter time *E*. Then Aristotle assumes that the speeds of the bodies are in the (inverse) ratio of the density of the two media. He goes on to conclude that a body would move infinitely rapidly through a void, since it would encounter no resistance. Because he considers this impossible, he deduces that a void or vacuum cannot exist.

In the paragraph just quoted, Aristotle attempts to strengthen his conclusion that there can be no void by arguing that, since a body moves more rapidly in direct proportion to its weight, bodies would move at different rates through a void without any apparent cause for this difference in speed. Rejecting this alternative, he states that all bodies would move at equal rates in a void, an equally unacceptable result for him. Hence, Aristotle rejects the possibility of the existence of a void. He holds that the medium is necessary for the motion of the body.

The Islamic philosopher Abu Muhammad Al-Farabi (850–970) appears to have carried out the first recorded experiments concerning the existence of vacuum, in which he investigated handheld plungers in water. He concluded that air's volume can expand to fill available space, and he suggested that the concept of perfect vacuum was therefore incoherent. In the Middle Ages, Christians held

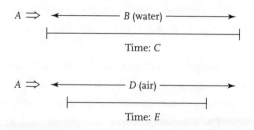

Fig 1. Diagram of the motion of a body (*A*) through a medium.

the idea of a vacuum to be immoral or even heretical, and it was not until 1643, when Evangelista Torricelli argued that there was a vacuum at the top of a mercury barometer, that the subject arose again.

Torricelli proposed that air has weight, and that it is the weight of air (not the attracting force of the vacuum, as assumed by Galileo) that holds (or rather, pushes) up the column of mercury settled vertically into a basin of mercury. He believed that the level at which the mercury stayed reflected the force of the air's weight pushing on it (specifically, pushing on the mercury in the basin and thus limiting the amount of mercury that can descend from the tube into the basin). In other words, he viewed the barometer as an instrument for measuring the pressure exerted by the atmosphere. The column's mercury height fluctuates with changing atmospheric pressure, as for example at different altitudes. Robert Boyle later conducted experiments on the properties of the vacuum. More precisely, he improved the first vacuum pump, which had been invented by Otto von Guericke in 1654, and conducted experiments on the properties of vacuum. He is also known for having conceived what has come to be called *Boyle's law*, which states that the volume of a gas varies inversely with the pressure exerted upon it.

Some people believe that although Torricelli produced the first sustained vacuum in a laboratory, it was Blaise Pascal who recognized it for what it was. In his *Expériences nouvelles touchant le vide* ("New Experiments with the Vacuum"), 1647, Pascal detailed basic rules describing to what degree various liquids could be supported by air pressure. He offered reasons to believe that there is indeed a vacuum above the column of liquid in a barometer tube. He also performed various experiments in Paris by which to demonstrate that the mercury drops to a lower level as atmospheric pressure rises. His insistence on the existence of the vacuum led to conflict with other prominent scientists, including Descartes.

Their barometer consisted of a tube approximately 84 cm in height, sealed at the top, filled with mercury, and set vertically into a basin of mercury at its base. The mercury in the tube adjusts until the weight of the mercury column balances the atmospheric force exerted on the basin. Higher atmospheric pressure exerts greater force on the basin, forcing the mercury higher in the column. Lower pressure allows the mercury to descend to a lower level in the column. Because higher temperature at the instrument will reduce the density of the mercury, the scale for reading the height of the mercury in the barometer is adjusted accordingly to compensate for this effect. What keeps the mercury suspended in the tube? Is there an unnatural vacuum that causes the surrounding glass to pull the liquid up? Or is there no vacuum at all but rather some rarefied and invisible matter in the "empty space"? Pascal answered that there really was nothing holding up the mercury. The mercury rises and falls in response to variations in the weight of the atmosphere. The mercury is being *pushed* up the tube, not pulled up by anything.

When Pascal offered this explanation to René Descartes, the latter wrote, rather nastily, in a letter to Christiaan Huygens, that Pascal has too much vacuum in his head. Descartes identified bodies with spatial extension, and so had no room for vacuums. If there were nothing between two objects, then they would be touching each other. And if they are touching each other, there is no gap between them.

Galileo was disappointed by Johannes Kepler's hypothesis that the Moon influences the tides, because the hypothesis seems to require causal chains in empty space. How could the great Kepler believe something so silly? When Isaac Newton resurrected Kepler's hypothesis, he was careful to suggest that the space between the Moon and the Earth is filled with ether. Indeed, the universality of Newton's laws of gravitation seems to require that the whole universe be filled with a subtle substance.

Newton's theory provoked many doubts. It is worth noting here that, in 1887, the Michelson-Morley experiment, using an interferometer, hopefully to detect change in the speed of light caused by the Earth's movement with respect to the ether, produced a famous null result, showing that there really is no static, pervasive medium throughout space. But although there is therefore no ether, and no such entity is required for the propagation of light, the space between the stars is not completely empty. Besides the various particles that constitute cosmic radiation, there is a cosmic background of photonic radiation (light), including the thermal background at about 2.7 K, seen as a relic of the Big Bang. But none of these findings affects the outcome of the Michelson-Morley experiment to any significant degree.

These doubts about the existence of ether were intensified by the emergence of Einstein's theory of relativity. Einstein argued that physical objects are not located in space, but rather have a spatial extent. Seen this way, the concept of empty space loses its meaning. Rather, space is an abstraction, based on the relationship between local objects. Nevertheless, the general theory of relativity admits a pervasive gravitational field, which, in Einstein's words, may be regarded as an "æther," with properties varying from one location to another. One must take care, however, not to ascribe to it material properties such as velocity and so on.

In 1930, Paul Dirac proposed a model of vacuum as an infinite sea of particles possessing negative energy, a model that came to be called the "Dirac sea." This theory helped refine the predictions of his earlier-formulated Dirac equation, and successfully predicted the existence of the positron, discovered two years later, in 1932. Dirac's idea was further developed in the context of quantum field theory.

Let us say a few more words about this striking idea. Dirac's electron, a 4-component entity, rather than one having just the two components of a "Pauli spinor," describes the two independent states of spin that a nonrelativistic particle of spin 1/2 possesses. In fact, Dirac's equation describes only two components of spin for a particle, despite there being four components for the wavefunction. Mathematically, the reason for this apparent disparity is closely related to the fact that the Dirac equation $\gamma^\mu \partial_\mu \psi = iM\psi$ is a first-order equation (we take $\hbar = 1$ in this expression), and its space of solutions is spanned by only half as many solutions as are spanned in the case of the second-order wave equation $(\Box + M^2)\psi = 0$. Physically, this "counting" of solutions of the Dirac equation must take into account the fact that the degrees of freedom of the electron's antiparticle, namely the positron, are also hiding in the solutions of the equation. It turns out that these solutions of the Dirac equation are *not* required to be of positive frequency, despite all of Dirac's cleverness and industry in eliminating the square root in the Hamiltonian. In fact, the presence of interactions, such as a background electromagnetic field, will cause an initially positive-frequency wave to gain negative-frequency parts.

When Dirac finally became convinced that the negative-frequency solutions could not be mathematically eliminated, he made the astounding suggestion that all the negative-energy states should already be fully occupied by the electrons (electrons satisfy the Pauli principle, and such a particle is not allowed to occupy a state if that state is already occupied). This ocean of occupied negative-energy states is now referred to as the "Dirac sea."

The core of Dirac's reasoning is thus the following. He proposed that almost all negative-energy quantum states of the electron in the universe are filled, already occupying all of the lower-energy states, so that, owing to the Pauli exclusion principle, no other electron could fall into them. The occasional unoccupied such state—a "hole" in this negative-energy sea—would appear as an antielectron (positron), thus having positive energy. So, then, an electron falling into such a hole would be interpreted as the annihilation of both electron and positron, which would produce a release of energy—the sum of the positive contributions from the electron and positron. Conversely, the supplying of sufficient energy to the Dirac sea could produce an electron-positron pair; in other words, if a hole were not present initially, but a sufficient amount of energy (normally in the form of photons) were to enter the system, then an electron could be ejected from one of the negative-energy states, thus leaving a hole. In this model, Dirac's "hole" is indeed the electron's antiparticle, now referred to as the *positron*.

The development of quantum mechanics, by requiring indeterminacy, has made more complex the modern interpretation of vacuum. Heisenberg's uncertainty principle (1927) predicts a fundamental uncertainty in the instantaneous measurability of the position and momentum of any particle, and thus, not unlike the gravitational field, questions the emptiness of space between particles. In the late twentieth century, especially with the development of the quantized field theory (QFT), the principle was understood to predict also a fundamental uncertainty in the number of particles in a region of space, leading to predictions of virtual particles arising spontaneously out of the void. In other words, there is a lower bound on the vacuum, dictated by the lowest possible energy state of the quantized fields in any region of space. In quantum mechanics, then, the *vacuum* is defined as the state (i.e., solution of the equations of the theory) having the lowest possible energy.

But even an ideal vacuum, thought of as the complete absence of anything, will not in practice be empty. One reason this is so is that the walls of a vacuum chamber emit light in the form of black-body radiation: visible light if the walls are at a temperature of thousands of degrees, infrared light if they are cooler. If this soup of photons is in thermodynamic equilibrium with the walls, the vacuum chamber can be said to have a particular temperature, as well as a particular pressure. Another reason why perfect vacuum is impossible lies in the Heisenberg uncertainty principle, which states that no particles can ever have an exact position. Each atom exists as a probability function on a Hilbert space, and thus space has a certain nonzero value everywhere in a given volume. Even the space between molecules is not a perfect vacuum.

More fundamentally, quantum mechanics predicts that the vacuum's energy will differ from its naïve, classical value. The quantum correction to the energy, called the *zero-point energy*, consists of the energies of virtual particles each of which has a brief existence. This condition, called *vacuum fluctuation*, may be related to the so-called cosmological constant of cosmology (see chapter 15 for relevant comments). The best evidence for vacuum fluctuations lies in the Casimir effect and the Lamb shift.

In QFT and string theory, the term "vacuum" is used to represent the ground state in the Hilbert space,[2] that is, the state with the lowest possible energy. In free (non-interacting) quantum field theories, this state is analogous to the ground state of a quantum harmonic oscillator. If the theory is obtained by the quantization of a classical theory, each stationary point of the energy in the configuration space gives rise to a single vacuum. String theory argues for a huge number of vacua—the so-called *string theory landscape* (see chapter 16 for a more thorough account of this topic).

Chapter 2

The Role of Vacuum
in Modern Physics

The vacuum is fast emerging as the central structure of modern physics. This issue is especially important in the context of classical gravity, quantum electrodynamics, and the grand unification program. The vacuum emerges as the synthesis of concepts of space, time, and matter; in the context of relativity and the quantum world, this new synthesis represents a structure of the most intricate and novel complexity. The synthesis raises genuine philosophical issues, and also longer links to modern metaphysics, in which the concepts of substance and space interweave in the most intangible of forms, the background and context of our experiences: vacuum, void, nothingness.

EMERGENCE OF THE CONCEPT OF QUANTUM VACUUM
IN COSMOLOGY

There is sufficient evidence at present to justify the belief that the universe began to exist without being caused to do so. This evidence includes the Hawking-Penrose singularity theorems, which are based on Einstein's general theory of relativity; the inflationary cosmological models of the early universe; and the quantum string models recently proposed by Gabriele Veneziano (1968) and other theoretical physicists. The Hawking-Penrose theorems led to an explanation of the beginning of the universe that involves the notion of a Big Bang singularity, whereas the quantum cosmological models represent the beginning largely in terms of the notion of a vacuum fluctuation. Theories that represent the universe as infinitely old, or as caused to begin, are shown to be at odds with—or at least unsupported by—these and other current cosmological notions.

It is nowadays recognized by most physicists that, in fact, the general theory of relativity (GTR) fails to apply to the physical world when quantum-mechanical interactions predominate, and these predominate when the temperature is at or above 10^{32} K, when the density is at or above 10^{94} gm cm^{-3}, and when the radius of curvature becomes on the order of 10^{-33} cm. Because of these conditions obtained during the Planck era, at the first 10^{-43} second after the singularity, the GTR-based Big Bang theory cannot be used as a reliable guide in reconstructing the physical processes that occurred during this time, and *a fortiori* cannot be used as a reliable basis for predicting that the density, temperature, and curvature had reached infinite values prior to this time.

Accordingly, it would seem that the foregoing probabilistic argument for an uncaused beginning of the universe is in difficulty. There are, however, three sound reasons for a continued support of the idea that the universe spontaneously began to exist. To comprehend the reasons, we must first observe that the reason why GTR is inapplicable during the Planck era is that the theory of gravity in GTR is unable to account for the quantum mechanical behavior of gravity during this era. A *new quantum theory of gravity* is needed. Although such a theory has not yet been developed, there are some general indications of what it may predict. It is in terms of these indications that our three reasons are to be understood.

First, it is thought that a quantum theory of gravity may show gravity to be repulsive rather than attractive, under the conditions that are obtained during the Planck era. During this time, regions of negative energy density may be created by the force and particles present, and these regions would lead to a gravitational repulsion. This suggests that any given finite set of past-directed time-like or null geodesics will not converge in a single point but will be pushed apart, as it were, by the repulsive gravitational force. This possibility is consistent with an oscillating universe, for as each contracting phase ends, gravity becomes repulsive and prevents converging geodesics from terminating in a point; gravity repels them, leading them to enter a new expanding phase.

But this way of avoiding the singularity predicted by the Hawking-Penrose theorems does not give us an infinitely old universe. For—and this is the first of the three reasons I want to mention—this oscillating quantum-gravitational universe would still be subject to some serious difficulties. One would be an increase in radius, cycle length, radiation, and entropy with each new cycle. Consequently, this theory does no more than others to push the cosmological singularity further into the past, at a time just before (or at) the beginning of the first cycle when the radius of the universe is zero (or near zero).

The second reason why Hawking-Penrose does not give us an infinitely old universe is that there is a way in which their theorems' prediction of a singularity at the outset of the present expansion can be made consistent with a quantum theory of repulsive gravity. These theorems do not *define* a singularity as that wherein curvature, density, and temperature are infinite and the radius is zero. A singularity is defined as a point or series of points beyond which the space-time manifold cannot be extended. Consequently, if the effects of quantum gravity prevent a buildup (an increase) of temperature, density, and curvature to infinite values, and a decrease of radius to zero, this need not be taken to mean that there is no singularity at the beginning of the present expansion. The singularity could occur with *finite* and *nonzero* values.

The third reason is that the most theoretically developed attempts to account for the past of the universe, on the basis of specifically quantum mechanical principles, have represented the universe as spontaneously beginning at the onset of the present expansion. These theories are collectively known as the "vacuum fluctuation models of the universe." The models developed by Tryon (1973), Brout, Englert, and Gunzig (1978), Grishchuk and Zeldovich (1982), Atkatz and Pagels (1982), and Gott (1982) picture the universe as emerging spontaneously from an empty background space, and the model of Vilenkin (also in 1982), which I shall discuss below, depicts it as emerging without cause from nothing at all.

In the first vacuum fluctuation model, developed by Edward Tryon, a vacuum fluctuation is an uncaused emergence of energy out of the empty space, an emergence governed by the *uncertainty relations* ("wave packets" can be described just as well in the momentum-space representation as they can in the position representation). One can introduce a precise notion of the "spread"—or lack of localization—of a wave packet in either the position description or the momentum description. Let us denote these spread measures, respectively, by Δx and Δp; Heisenberg's uncertainty relation tells us that the product of these spreads cannot be smaller than on the order of Planck's constant.

Position states, momentum states, and wave packets may be depicted in both the momentum and position representations. We have it that, in the case of a pure momentum state, the spread in the momentum is zero, so $\Delta p = 0$ (i.e., a delta function in momentum space). From the Heisenberg relation, Δx is now infinite, in accordance with the condition that the wave-function becomes spread uniformly over the whole of position space. The situation is just the opposite with a position state, where now $\Delta x = 0$, the position being defined with complete precision, but where the spread Δp in the momentum state now becomes infinite.

It is interesting to see that here we have examples that clearly illustrate the incompatibility of noncommuting measurements in quantum mechanics. A measurement of a particle's momentum would put it into a momentum state, corresponding to some classical value P, and any subsequent measurement of the momentum in this state would yield the same result P. But if the state were instead subjected to a subsequent *position* measurement, following an initial measurement of momentum, the result would be completely uncertain, and any one result for the position would be as likely as any other. This measurement makes the momentum state a delta function in position. In momentum space, this state is a plane wave, spread out uniformly in all possible values for the momentum. A subsequent *momentum* measurement would then be completely uncertain. Thus, the very act of intermediate position measurement has completely ruined the purity of the original momentum state.

Let's also mention that, consistent with relativity, there is a similar Heisenberg uncertainty relation between energy and time; the usual interpretation of the energy/time uncertainty is that if the energy of a quantum system is ascertained in some measurement performed in a time Δt, then there is a ΔE uncertainty in this energy measurement that must satisfy the relation.[1]

$$\Delta x \Delta p \geq \hbar/2$$
$$\Delta t \Delta E \geq \hbar/2,$$

and thus has zero net value for conserved quantities. (Recall that Heisenberg's principle of uncertainty states that the position and momentum of a particle cannot be measured simultaneously with arbitrarily high precision. There is a minimum for the product of the uncertainties of these two measurements. There is likewise a minimum for the product of the uncertainties of energy and time.[2]) Tryon (1973) argued that the universe is able to be a fluctuation from a vacuum in the larger space in which the universe is embedded, since it does have a zero net value for its conserved quantities. Observational evidence (Tryon claims) supports or is consistent with the fact that the positive mass-energy of the universe is canceled by its negative gravitational potential energy, and that the amount of matter created is equal to the amount of antimatter.

A disadvantage of Tryon's theory, and of other theories that postulate a background space from which the universe fluctuates, is that they explain the existence of the universe but only at the price of introducing another unexplained given, namely, the background space. This problem is absent from Vilenkin's theory (1982), which represents the universe as emerging without a cause "from literally *nothing*" (see chapter 15 for a more detailed account of this theory). The universe appears in a quantum tunneling from nothing at all to de Sitter space. Quantum tunneling is normally understood in terms of processes *within* space-time; an electron, for example, tunnels through some barrier even though it lacks sufficient energy to cross it but nevertheless does cross. This is possible because the above-mentioned uncertainty relation allows the electron to spontaneously acquire additional energy for the short period of time required for it to tunnel through the barrier. Vilenkin applies this concept to space-time itself; in this case, there is not a state of the system before the tunneling, for the tunneling is the first state that exists. The state of tunneling is thus an analogue of the Big Bang, according to the definition that the universe began neither at nor after the singularity, for the tunneling itself is the first state of the universe, and there *is* no time previous to this state. The equation describing this state is a quantum-tunneling equation, specifically the bounce solution of the Euclidean version of the evolutionary equation of a universe with a closed Robertson-Walker metric (1935, 1937).[3] The universe emerged from the tunneling with a finite size ($a = cH^{-1}$) and with a zero rate of expansion or contraction ($da/dt = 0$). It emerged in a symmetrical vacuum state, which then decays and the inflationary era begins; when this era ends, the universe evolves according to the standard Big Bang model.

These quantum-mechanical models of the beginning of the universe are explanatorily superior in one respect to the standard GTR-based Big Bang models: they do not postulate initial states at which the laws of physics break down, but rather explain the beginning of the universe *in accordance with* the laws of physics. The GTR-based theory, by contrast, predicts a beginning of the universe by predicting initial states at which the laws of the theory that are used to predict these states break down. The singularity and the explosion of four-dimensional space-time emerging from the singularity obey none of the laws of GTR that are obeyed by states within the universe or by subsequent states of the universe. In contrast, the quantum-mechanical theories represent the universe as coming into existence via the same laws that processes within the universe obey. Instead of an exploding singularity, there is a quantum fluctuation or tunneling that is analogous to the fluctuations or tunnelings within the universe, and obeys the same uncausal laws as the latter fluctuations or tunnelings.

The previous issue—that of the quantum fluctuation or tunneling of the early universe—is closely related to the subject of quantum cosmology, and particularly to the path-integral approach for cosmology. As proposed by Hartle and Hawking in 1983, this path integral is to be taken over all compact metrics without boundary. Let us first suppose that it is to be taken over all asymptotically Euclidean metrics. Then there would be two contributions to probabilities for measurements in a finite region (inside the universe). The first contribution would be from connected asymptotically Euclidean metrics. The second would be from two disconnected metrics, one consisting of a compact space-time containing the region of measurements and the second a separate asymptotically Euclidean metric.

One cannot exclude disconnected metrics from the path integral, because they can be approximated by connected metrics in which the different components are joined by thin tubes or wormholes of negligible action. Disconnected compact regions of space-time won't affect scattering calculations, because they aren't connected to infinity, where all measurements are made. But they will affect measurements in cosmology that are made in a finite region. Indeed, the contributions from such disconnected metrics will dominate the contributions from connected asymptotically Euclidean metrics. Thus, even if one took the path integral for cosmology to be dominant to all asymptotically Euclidean metrics, the effect would be almost the same as if the path integral had been over all compact metrics.

Let us then suppose that the path integral for quantum gravity is taken over all compact metrics without boundary; call it the *Hartle and Hawking proposal*. Then, we need to define geometrically the state of the universe at one time. Consider the probability that the space-time manifold M contains an embedded three-dimensional manifold Σ with induced metric h_{ij}. This is given by a path integral over all metrics g_{ab} on M that induce h_{ij} on Σ. Then one defines the probability of induced metric h_{ij} on Σ as

(2.1) $$P\, h_{ij}(\Sigma) : \int_{g_{ab}(M) \to h_{ij}\,\Sigma} d[g]e^{-1}.$$

Assume M to be simply connected. Then the surface Σ will divide M into two parts, M^+ and M^- (fig. 2). In this case, the probability that Σ can have the metric h_{ij} can be factorized. It is the product of two wave functions Ψ^+ and Ψ^-. These are given path integrals over all metrics on M^+ and M^-, respectively, that induce the given three-metric h_{ij} on Σ. So we have

(2.2) $$P(h_{ij}) = \Psi^+(h_{ij}) \times \Psi^-(h_{ij}), \text{ where}$$
$$\Psi\, h_{ij}(\Sigma) : \int_{g_{ab}(M) \to h_{ij}(\Sigma)} d[g]e^{-1}.$$

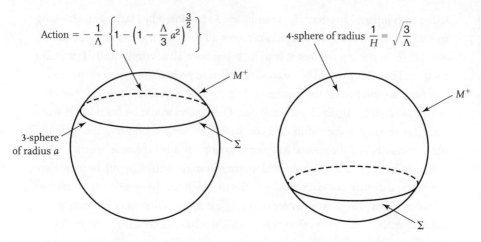

Fig. 2. The two possible Euclidean solutions M^+ with boundary Σ, and their actions.

In most cases, the two wave functions will be equal. Hence, we can drop the superscripts $+$ and $-$. Ψ is called the wave function of the universe. If there are matter fields ϕ, the wave function will also depend on their values ϕ_0 on Σ. But it will not depend explicitly on time, because there is no preferred time coordinate in a closed universe. The no-boundary proposal implies that the wave function of the universe is given by a path integral over fields on a compact manifold M whose only boundary is the surface Σ. The path integral is taken over all metrics and matter fields on M that agree with the metric h_{ij} and the matter fields ϕ_0 on Σ.

One can describe the position of the surface Σ by a function τ of three coordinates x_i on Σ. But the wave function defined by the path integral cannot depend on τ, or on the choice of the coordinates x_i. This implies that the wave function Ψ must obey four functional differential equations. Three of these equations, called the *momentum constraints*, are written as follows: $(\partial \Psi / \partial h_{ij}) = 0$. They express the fact that the wave function should be the same for different three-metric h_{ij} that can be obtained from each other by transformations of the coordinates x_i. The fourth equation, called the *Wheeler-DeWitt equation*, given as

(2.3) $$(G_{ijkl}(\partial^2 / \partial h_{ij} \, \partial h_{kl}) - h^{1/2} \, {}^3R) \, \Psi = 0,$$

corresponds to the independence of the wave function of τ. One can think of it as the Schrödinger equation for the universe. But there is no time-derivative term, because the wave function does not depend on time explicitly.

In order to estimate the wave function of the universe, one can use the saddle point approximation to the path integral, as in the case of black holes. One finds a Euclidean metrics g_0 on the manifold M^+ that satisfies the field equations and induces the metric h_{ij} on the boundary Σ. One can then expand the action in a power series around the background metric g_0. Thus,

(2.4) $$I[g] = I[g_0] + 1/2\delta g I_2 \delta g + \ldots$$

As before, the term "linear" in the perturbations vanishes. The quadratic term can be regarded as giving the contribution of gravitons on the background, and the higher-order terms as interactions between the gravitons. These can be ignored when the radius of curvature of the background is large compared to the Planck scale. Therefore,

(2.5) $$\Psi \approx 1/(\det I_2)^{1/2} e^{-I[g_0]}.$$

From a simple example, one can see what the wave function is like. Consider a situation in which there are no matter fields, but there is a positive cosmological constant Λ. Let us take the surface Σ to be a three-sphere and the metric h_{ij} to be the round three-sphere metric of radius a. Then the manifold M^+ bounded by Σ can be taken to be the four-ball. The metric that satisfies the field equations is part of a four-sphere of radius $1/H$, where $H^2 = \Lambda/3$. The action is

(2.6) $$S_E = 1/16\pi \int (R - 2\Lambda) \, (-g)^{1/2} \, d^4x + 1/8\pi \int K(\pm h)^{1/2} \, d^3 x.$$

For a three-sphere Σ of radius less than $1/H$, there are two possible Euclidean solutions: M^+ can be less than a hemisphere or it can be more (fig. 2). But there are arguments showing that one should choose the solution corresponding to less than a hemisphere. The next figure (fig. 3) shows the contribution to the wave function that arises from the action of the metric g_0. When the radius of Σ is less than $1/H$, the wave function increases exponentially like $\exp(a^2)$, where a is the radius of a round three-sphere with metric h_{ij}. But when a is greater than $1/H$, one can analytically continue the result for smaller a and obtain a wave function that oscillates very rapidly.

One can interpret this wave function as follows. The real-time solution of the Einstein equations with an Λ term and maximal symmetry is de Sitter space. This can be embedded as a hyperboloid in 5-dimensional Minkowskian space, and the Lorentzian de Sitter metric is then

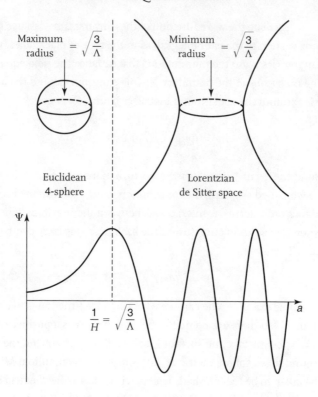

Fig. 3. The wave function as a function of the radius of Σ.

(2.7) $ds^2 = -dt^2 + 1/H^2 \cosh Ht(dr^2 + \sin^2 r(d\theta^2 + \sin^2 \theta d\phi^2)).$

One can think of this metric as a closed universe that shrinks down from infinite size to a minimum radius and then expands again exponentially. The metric can be written in the form of a Friedmann universe with scale factor cos hHt. Putting $\tau = it$ converts the cos h into cos, giving the Euclidean metric on a four-sphere of radius $1/H$. Thus,

(2.8) $ds^2 = d\tau^2 + 1/H^2 \cos H\tau(dr^2 + \sin^2 r(d\theta^2 + \sin^2 \theta d\phi^2)).$

Thus, one gets the idea that a wave function that varies exponentially with the three-metric h_{ij} corresponds to an imaginary-time Euclidean metric. On the other hand, a wave function that oscillates rapidly corresponds to a real-time Lorentzian metric.

As in the case of the pair creation of black holes, one can describe the spontaneous creation of an exponentially expanding universe. One joins the lower half of the Euclidean four-sphere to the upper half of the Lorentzian hyperboloid (fig. 3). Unlike what can be said about the black hole pair creation, one cannot say that the de Sitter universe was created out of field energy in a preexisting space. Instead, it would quite literally be created out of nothing, not just out of the vacuum, but out of absolutely nothing at all, because there is nothing outside the universe (see chapter 15 for a more detailed account of this subject). In the Euclidean regime, the de Sitter universe is just a closed space, like the surface of the Earth, but with two additional dimensions. If the cosmological constant is small compared to the Planck value, the curvature of the Euclidean four-sphere should be small. This will mean that the saddle-point approximation to the path integral should be good, and that the calculation of the wave function of the universe will not be affected by our unawareness of what happens at very high curvatures.

We should here recall that the de Sitter space, denoted dS_n, is the Lorentzian analog of an n-sphere (with its canonical Riemannian metric). It is a maximally symmetric Lorentzian manifold with constant positive curvature, and is simply connected for $n \geq 3$. In the language of general relativity, de Sitter space is the maximally symmetrical vacuum solution of Einstein's field equation with a positive (repulsive) cosmological constant Λ. When $n = 4$, de Sitter space is also a cosmological model for the physical universe.

A de Sitter universe has no ordinary matter content but does have a positive cosmological constant that sets the expansion rate, H. A larger cosmological constant leads to a larger expansion rate: $H \propto \sqrt{\Lambda/M_{PL}}$, where the constant of proportionality depends on conventions. The cosmological constant is Λ, and M_{PL} is the Planck mass. It is common to describe a patch of this solution as an expanding universe of the FLRW form, where the scale factor is given by $a(t) = e^{Ht}$, where the constant H is the Hubble expansion rate and t is time. As in all FLRW spaces, the scale factor $a(t)$ describes the expansion of physical spatial distances. Our universe may be approaching a de Sitter universe in the infinite future. If the current acceleration of our universe is due to a cosmological constant, then, as the universe continues to expand, all of the matter and radiation will be diluted. Eventually, there will be almost nothing left but the cosmological constant, and our universe will have become a de Sitter universe. The exponential expansion of the scale factor means that the physical distance between any two non-accelerating observers will eventually be increasing faster than the speed of light. At this point, those two observers will no longer be able to make contact. Therefore, any observer in a de Sitter universe would see event horizons beyond which that observer can never see nor learn any information. Another application of de Sitter space is to be found in the early universe during cosmic inflation. Many inflationary models are approximately de Sitter space and can be modeled by giving the Hubble parameter a mild time dependence.

Mathematically, de Sitter space can be defined as a submanifold of Minkowskian space in the five-dimension space $\mathbb{R}^{1,4}$. Take Minkowskian space $\mathbb{R}^{1,n}$ with the standard metric

(2.9) $$ds^2 = -dx_0^2 + \sum_{i=1}^{n} dx_i^2.$$

De Sitter space is the submanifold described by the hyperboloid

(2.10) $$-x_0^2 + \sum_{i=1}^{n} x_i^2 = \alpha^2,$$

where α is some positive constant with dimensions of length. The metric on de Sitter space is the metric induced from the ambient Minkowski metric. One can confirm that the induced metric is nondegenerate and has Lorentzian signature. De Sitter space can also be defined as the quotient $O(1, n)/O(1, n-1)$ of two indefinite orthogonal groups, which shows that it is a non-Riemannian symmetric space. Topologically, de Sitter space is $\mathbb{R} \times S^{n-1}$ (so that if $n \geq 3$, then de Sitter space is simply connected). The isometry group of de Sitter space is the Lorentz group $O(1, n)$. The metric therefore has $n(n + 1)/2$ independent Killing vectors and is maximally symmetric. Every maximally symmetric space has constant curvature. The Riemannian curvature tensor of de Sitter space is given by

(2.11) $$R_{\rho\sigma\mu\nu} = 1/\alpha^2(g_{\rho\mu}g_{\sigma\nu} - g_{\rho\nu}g_{\sigma\mu}).$$

De Sitter space is an Einstein manifold, since the Ricci tensor is proportional to the metric: $R_{\mu\nu} = (n - 1/\alpha^2)\, g_{\mu\nu}$. This means that de Sitter space is a vacuum solution of Einstein's equation, with the cosmological constant given by

(2.12) $$\Lambda = (n - 1)(n - 2)/2\alpha^2.$$

The scalar curvature of de Sitter space is given by

(2.13) $$R = n(n - 1)/\alpha^2 = (2n/n - 2)\Lambda.$$

For the case $n = 4$, we have $\Lambda = 3/\alpha^2$ and $R = 4\Lambda = 12/\alpha^2$.

NON-ABELIAN YANG-MILLS THEORY

Beginning in the 1970s, 20 years after the discovery by Chen Ning Yang and Robert Mills (1954) of a non-Abelian gauge theory for strong force (nuclear interactions) in which the local gauge group was the $SU(2)$ isotopic-spin group,[4] these physicists were able to express the concept of a gauge field in such a way that it could be recognized as an instance of more abstract structures known to mathematicians as connections in fiber bundles. The discovery of this equivalence has made it possible to understand that mathematical concepts and structures are theoretically powerful and therefore both necessary and suitable for the description and explanation of physical reality.

Precisely, the mathematical structure of gauge theory is that of a vector bundle E with structure group G over a compact Riemannian manifold M. We assume that $G \subset O(m)$, and that E carries an inner product compatible with G. Let $\mathcal{E}$ be the space of G-connections on E, and let $\mathcal{G}$ be the space of G-automorphisms of E. Then $\mathcal{G}$ acts on $\mathcal{E}$ as before, and we have a quotient space $\mathcal{B} \equiv \mathcal{E} / \mathcal{G}$. To each connection $\nabla \in \mathcal{E}$, there is associated a curvature 2-form R^∇, and at each point x we can take its norm to be

$$(2.14) \qquad \|R^\nabla\|_x^2 \equiv \sum_{i<j} \|R^\nabla_{e_i, e_j}\|_x^2,$$

where $\{e_1, \ldots, e_n\}$ is an orthonormal basis of $T_x M$, and the norm of $R^\nabla_{e_i, e_j}$ is the usual one on $\mathrm{Hom}(E, E)$—namely, $\langle A, B \rangle \equiv \mathrm{Trace}\,(A^t \circ B)$. Given any $g \in \mathcal{G}$, we recall that $R^{g(\nabla)} = g \circ R^\nabla \circ g^{-1}$, so that

$$(2.15) \qquad \|R^{g(\nabla)}\| \equiv \|R^\nabla\| \text{ on } M.$$

This says that the pointwise norm of the curvature is gauge-invariant.

The mathematical basis of gauge field theory lies in vector bundles and the connections on them. One of the most striking developments in mathematical physics over the past quarter century has been the discovery of the fundamental role played by bundles, connections, and curvature in expressing and eventually explaining the basic laws of nature. The so-called Yang-Mills theory does reflect, in a deeper way, the intimate relationship between geometrical concepts and physical ideas. The key feature of Yang-Mills theory is the invariance of the physical properties of particles under an infinite-dimensional group G, the non-Abelian gauge group $SU(2)$. The classical Maxwell's equations of 1865 in vacuum for the electromagnetic field are invariant under the $U(1)$ gauge group.

More precisely, both Abelian and non-Abelian theories—Maxwell's theory and the Yang-Mills theory—are invariant under infinite parameter groups, $U(1)$ and $SU(2)$ respectively. They are also both invariant under Lorentz transformations. Quantum chromodynamics is a non-Abelian gauge theory[5] with the action of the $SU(3)$ group on the color triplet of quarks.

Recall briefly the mathematical setting for gauge-field theory. The configuration space of the theory is taken to be the space $\mathcal{A}_{P(M,G)}$ of all gauge potentials (i.e., connections) on the principal bundle $P(M, G)$. The (classical) gauge field is the curvature of a connection on the bundle $P(M, G)$. The structure group G is called the *gauge group*. The group G_P of automorphisms of the bundle P covering the identity is called the *group of gauge transformations*. The Lagrangian is defined as a function of the configuration space, and the corresponding quantum field theory is constructed by considering the space of classical fields as a configuration space C and defining the quantum expectation values of gauge invariant functions on C by using path integrals. This is usually referred to as the Feynman path integral method of quantization. The starting point of this method is the choice of a Lagrangian defined on the configuration space of classical gauge fields. This Lagrangian is used to define the action functional that emerges in the integrand of the Feynman path integral.

Let us add a historical note. The origin of Feynman's path integral concept can be found in Dirac's important book *The Principles of Quantum Mechanics* (1930), wherein the author introduced the notation $e^{iS/\hbar}$. Here, Dirac's S is the familiar action quantity $\int L \, dt$, where $L(x, \dot{x}, t)$ is the Lagrangian of classical mechanics. Now, how does this come about? The Lagrangian has everything built into it—kinetic and potential energy, including interaction terms—so that if Dirac's remark is true, then the concept of *propagator* (used by Schrödinger and Heisenberg in quantum dynamics) is truly a wonderful discovery. Could it be that a principle of least action holds in quantum mechanics (as it does in classical mechanics), such that by minimizing S the quantity $\langle x_n, t_n | x_0, t_0 \rangle$ will describe the true path of a particle? Feynman had been fascinated by this idea of Dirac's. How does the Lagrangian enter into it, anyway? Feynman made a long story short, took the idea, and made it the basis of his Ph.D. dissertation, "The Principle of Least Action in Quantum Mechanics," and in so doing discovered a completely new approach to quantum theory—the *path integral*. You can do it, too. Just recognize that $\langle x_n, t_n | x_0, t_0 \rangle = \langle x_n | U(t_n - t_0) | x \rangle$, and remember that U contains the Hamiltonian operator H, which provides all the physics.

Extracting the $e^{iS/\hbar}$ term from all this is just one step in what all physicists consider to be one of the most profound discoveries of quantum physics—the path integral approach to quantum field theory. The quantity $\langle x_n, t_n | x_0, t_0 \rangle$ represents an infinite set of paths that a particle can take from one point to another over a finite time. But in classical mechanics there is only one path that the particle can take, the so-called classical path. Although it is also defined by a Lagrangian, the difference between one path and an infinite number of paths is obviously quite confusing. In the classical scheme, Planck's constant $\hbar$ is, for all practical purposes, zero; but when set at $\hbar = 0$, the quantity S in the path integral oscillates violently and becomes meaningless.

There is only one way out of this. If the Lagrangian S is also set to zero, then in the limit the indefinite quantity $S/\hbar = 0/0$ might somehow leave something finite behind. Unfortunately, although S may have some minimum value, setting it exactly to zero is usually not valid. Nevertheless, minimizing S is precisely the route one takes to get onto the classical path, and this is the only option available to us. One way of looking at this is that even in the classical world, $\hbar$ is not exactly zero either, so the ratio remains finite. This fact singles out one unique path, which is the classical path. Remarkably, at the quantum level there is no unique path—all possible paths contribute to the transition amplitude. And even more amazingly, each path is just important as any other. It is only when $\hbar$ is comparatively small that the paths begin to interfere destructively, leaving a large propagation amplitude only in the vicinity of the classical path.

When you studied the electron double-slit experiment, you learned that each electron in reality passes both slits on the screen on its way to the detector and, in doing so, interferes with itself, which is why the detector shows an interference pattern. For a triple slit, the electron has three possible routes, and there is a corresponding interference pattern. We can in fact make an infinite number of slits in an infinite number of sequential screens (leaving empty space!), and the electron will then be describable by Feynman's path integral formalism. Fantastic as this may seem, the formalism appears to be a correct description of reality.

QUANTUM FIELD THEORIES

A quantum field theory may be considered as an assignment of the quantum expectation $\langle \Phi \rangle_\mu$ to each gauge invariant function $\Phi : \mathcal{A}(M) \to \mathbf{R}$. A gauge invariant function $\Phi : \mathcal{A}(M) \to \mathbf{R}$ is called an observable in quantum field theory. In the Feynman path integral approach to quantization, the quantum expectation $\langle\Phi\rangle_\mu$ of an observable is given by the expression

$$(2.16) \qquad \langle\Phi\rangle_\mu = \int_{\mathcal{A}(M)} e^{-S_\mu(\omega)} \Phi(\omega) \mathcal{D}\mathcal{A} / \int_{\mathcal{A}(M)} e^{-S_\mu(\omega)} \mathcal{D}\mathcal{A},$$

where $\mathcal{D}\mathcal{A}$ is a suitably defined measure on $\mathcal{A}(M)$. It is usual to express the quantum expectation $\langle\Phi\rangle_\mu$ in terms of the partition function Z_μ, defined by

$$(2.17) \qquad Z_\mu(\Phi) := \int_{\mathcal{A}(M)} e^{-S_\mu(\omega)} \Phi(\omega) \mathcal{D}\mathcal{A}.$$

Thus, we can write

$$(2.18) \qquad \langle\Phi\rangle_\mu = Z_\mu(\Phi)/Z_\mu(1).$$

In the above equations we have written the quantum $\langle\Phi\rangle_\mu$ to indicate explicitly that we have, in fact, a one-parameter family of quantum expectations indexed by the coupling constant μ in the action.

There are several examples of gauge invariant functions. For example, primary characteristic classes evaluated on suitable homology cycles give an important family of gauge invariant functions. The instanton number k of $P(M, G)$ belongs to this family, since it corresponds to the second Chern class evaluated on the fundamental cycle of M representing the fundamental class $[M]$. The pointwise norm $|F_\omega|_x$ of the gauge field at $x \in M$, the absolute value $|k|$ of the instanton number k, and the Yang-Mills action are also gauge invariant functions. Another important example of a quantum observable is given by the Wilson loop functional $\mathcal{W}_{\rho,\alpha}$ associated with the representation ρ and the loop α. The definition of the Wilson loop functional and a given equation expressing a gauge transformation, which changes the holonomy by conjugation by an element of the gauge group G, implies that the Wilson loop functional is gauge invariant and hence defines a quantum observable. Regarding a knot κ as a loop, we obtain a quantum observable $\mathcal{W}_{\rho,\kappa}$ associated with the knot.

It might be useful to recall that the concept of *holonomy* originated between 1920 and 1950 in the theory of connections of differentiable fiber bundles (T. Levi-Civita, H. Weyl, E. Cartan, and C. Ehresmann). Let $P = (P, \pi, M, G)$ be a differentiable principal bundle. (We assume that differentiability always means that of class C^∞.) The total space P and the base space M are differentiable manifolds, and the projection π is a differentiable mapping. The structure group G, a

Lie group, acts on P from the right as a transformation group. On each fiber, G acts transitively without fixed points. For elements a, x in G, P, we write $R_a(x) = xa$. The mapping induced on tangent vector spaces by R_a and π will be denoted by the same letters, namely $R_a : T_x(P) \to T_{xa}(P)$, $\pi : T_x(P) \to T_{\pi(x)}(M)$. The tangent vector space $T_x(P)$ at each point x of P is mapped by the projection π onto the tangent vector space $T_p(M)$ at the point $p = \pi(x)$ of M. The kernel of this mapping is denoted by $V_x(P)$, and each vector in $V_x(P)$ is said to be *vertical*. The kernel $V_x(P)$ is the totality of elements of $T_x(P)$ that are tangent to the fiber. We say that a *connection* is given in P if for each point $x \in P$, a subspace Q_x of the tangent space $T_x(P)$ is given in such a way that the following three conditions are satisfied: (i) $T_x(P) = V_x(P) + Q_x$ (direct sum); (ii) $R_a(Q_x) = Q_{xa}$ (Q is invariant under G); and (iii) the mapping $x \to Q_x$ is differentiable. A vector in Q_x is said to be *horizontal*. Now suppose that X is an arbitrary vector field on P. By condition (i), the value X_x of X at each point x of P can be expressed uniquely as $X_x = Y_x + Z_x$, where $Y_x \in V_x(P)$ and $Z_x \in Q_x$. The vector fields Y and Z defined by Y_x and Z_x ($x \in P$) are called the *vertical* and *horizontal* components of X, respectively. Condition (iii) implies that if X is a differentiable vector field, then its horizontal and vertical components are also differentiable vector fields. Let X be a vector field on the base space M. Since π defines an isomorphism of Q_x and $T_p(M)$ ($p = \pi(x)$), we have a unique vector field X^* on P such that (a) $\pi(X^*) = X$ and (b) $X_x^* \in Q_x$. We call X^* the *lift* of X, and it is invariant under G by condition (ii).

Suppose that a connection is given in P. If C is a piecewise differentiable curve in the base space M, we can define a mapping φ that maps the fiber over the initial point p of C onto the fiber over the endpoint q of C as follows: Take an arbitrary point x on the fiber at p. Then we have a unique curve C_x^* in P starting at x such that (a) $\pi(C_x^*) = C$, and (b) each tangent vector to C_x^* is horizontal. (C_x^* is called a *lift* of C that starts at x.) The endpoint y of the curve C_x^* belongs to the fiber over q. We set $\varphi(x) = y$. Because $C_{xa}^* = R_a(C_x^*)$, the mapping φ commutes with transformations of G. We call this mapping φ the *parallel displacement* or *parallel translation* along the curve C. Now, fix a point p in the base space M. If C is a closed curve in M starting from p, the parallel displacement along C maps the fiber over p onto itself. So if we fix a point x on the fiber over p, x is transformed by the parallel displacement to a point xa ($a \in G$). Thus each closed curve C starting from p determines an element $a(x, C)$ of G. If C varies over the set of closed curves that start from p, the totality of such elements of G forms a subgroup of G. This subgroup is called the *holonomy group* of the connection defined over P with the reference point x. If M is connected, holonomy groups with different reference points are conjugate. In the above, if we choose as the closed curves C starting from p only those curves that are null-homotopic, the elements $a(x, C)$ form a subgroup of the holonomy group. This is called the *restricted holonomy group*. The holonomy group is a Lie subgroup of the structure group G, and its connected component containing the identity coincides with the restricted holonomy group. Holonomy groups are useful in the study of the behavior of connections.

Chapter 3

The Quantum Vacuum in the Early Universe

Quantum mechanics was never ignored in the development of cosmology. The properties of matter and radiation, spectral lines, light scattering, the statistics of Bose or Fermi—all these topics were taken into account in the calculation of pressure, energy density, spectrum transport coefficients, etc. Therefore, the right-hand side of Einstein's general relativity equations

$$(3.1) \qquad R_{\mu\nu} - 1/2 g_{\mu\nu} R + \Lambda g_{\mu\nu} = 8\pi G \, T_{\mu\nu}$$

already included quantum effects. Speaking about these effects now, we emphasize the influence of space-time curvature on particles and fields, as opposed to the usual physics of Minkowskian space.

The most interesting effect is the creation of particles by the gravitational field in vacuum. The reactions of the type $e^+ + e^- = g + g$, where g is gravitons, were considered and calculated in the 1930s and 1940s. Taking many coherent gravitons, we obtain a classical gravitational wave. The creation of e^+e^- pairs obviously occurs in colliding beams of (classical) gravitational waves, i.e., in a vacuum with time-dependent metric.

In a cosmological context, the excitation of fields, i.e., the creation of field quanta (for example photons) in an expanding universe, was mentioned by E. Schrödinger (1939[1], also 1956), and later by R. Utiyama and B. S. DeWitt (1962). A thorough investigation was made by L. Parker (1968, 1969). The general principle is that particle creation is due to the non-adiabatic behavior of the corresponding field in a changing metric. The particles are created with a frequency of the order of the inverse characteristic time of the change of metric. This principle leads to the threshold of massive particle creation: no creation at $t > h/mc^2$. By dimensional arguments, the energy density of created particles is of the order of h/c^3t^4. The important qualitative features of particle creation are two:

The first is the vacuum polarization effect (see below)—the appearance of energy-stress tensor components without real particles. One example is the Casimir effect (1948; see below) at zero temperature in the static situation. Here, pure vacuum polarization occurs. In the case of a small localized perturbation, $g_{\mu\nu} = \eta_{\mu\nu} + h_{\mu\nu}$, where $h_{\mu\nu} = 0$ at $t < t_1$ and $t > t_2$, $h_{\mu\nu} \ll 1$ and smooth in the interval $t_1 < t < t_2$. The vacuum polarization is characteristically of first order in $h_{\mu\nu}$ and vanishes together with the metric perturbation for $t > t_2$.

The second important feature of the theory is its conformal invariance—at least in the limit of vanishing rest-masses. The isotropic expansion is conformally equivalent to a static situation. Therefore, particles creation is abnormally small (proportional to $Gm^2/hc = 10^{-38}$ in some positive power) in Friedmann isotropic and homogeneous cosmological models. On the other hand, creation of particles in an anisotropic expansion gives divergent results if switched on at $t = 0$. If one (artificially) turns on the creation at $t = t_{pl} = 10^{-43}$ s, it destroys the anisotropy during a time of the same order. From these considerations it follows that cosmological solutions, which are anisotropic at the singularity, are prohibited. But we are not left solely with the Friedmann models. The quasi-isotropic solutions with isotropic expansion in the limit $t \to 0$ are also possible. They have an inhomogeneous three-space metric. Therefore, density perturbations as well gravitational waves evolve from the quasi-isotropic solution. At the same time, anisotropy also emerges—but sufficiently late that particle creation could be avoided if the three-space metric inhomogeneities are smooth enough.

The most important and most beautiful result of the marriage between quantum theory and gravity is Stephen Hawking's (1975) theory of black hole evaporation. Concerning its physical content, one can say, in a simplified manner, that collapse to a black hole is an event that does not end in a static situation (unlike the collapse of a white dwarf to a neutron star, for example). The black hole formation leads to the exponential reddening $\omega \sim e^{-ct/r_g}$, of all outgoing radiation. Therefore, at all times the reddening is non-adiabatic for outgoing waves with frequency $\omega \sim c/r_g$, and the collapsing black hole thus emits particles with this frequency. This order of magnitude argument of Hawking was substantiated by arguments proving the detailed character of the radiation.

The evaporation is not important for stellar black holes, but only for primordial ones. The idea of primordial black holes (PBH) was first formulated by Zel'dovich and Novikov (1967).

If the initial metric and density distribution near the singularity are sufficiently inhomogeneous, it is quite possible that in some parts of the primordial plasma the expansion can be followed by compression ending in early gravitational collapse, i.e., the formation of PBH. It must be stressed that the PBH formation is a phenomenon beyond the scope of the linear theory of perturbations, because in this theory the small-scale perturbations develop into oscillations and are totally damped very early, thus leaving no observable traces.

Let us now ask, "How have our views of black holes changed since the discovery by Hawking of PBH quantum evaporation?" If the mass of the PBH is greater than the Planckian mass $(\hbar c/G)^{1/2} = 10^{-5}\,g$, i.e. if the PBH collapse occurs later than the Planckian time, $10^{-43}\,s$, then the formation and evaporation are separated in time, and the evaporation does not affect formation. The link between the initial metric perturbations and PBH formation thus remains. Detection methods and an initial mass on the order of 3×10^{14} to $10^{16}\,g$ are still alive or died not long previously, and they are (or were) strong emitters of hard γ photons, electron-positron pairs, and mesons. Such highly active objects would constitute a major contribution to cosmic rays, x-rays, and γ-ray background. In the mass range mentioned, the detection sensitivity of active (evaporating) PBHs is 10^8 times greater than the sensitivity of passive PBH detection by their gravitation, and the observable limit for evaporating PBHs is much lower.

A PBH with evaporation times in the interval $100\,s < t < 10^5$ years (before recombination) would spoil the Planckian spectrum of the background radiation. PBHs that evaporate earlier $(1s < t < 100\,s)$ change the composition of nucleosynthesis. At $t < 1s$, a PBH will lead to an entropy increase—a part of entropy may actually be due to this process—but one must not obtain more than the observed entropy per baryon, i.e., the contribution of PBH must be less than 10^9 dimensionless entropy units per baryon.

In the context of evaporation theory, one can suppose that the abundant initial formation of PBH followed the very early evaporation. Strong departure from homogeneity is possible on the smallest scale—from Planckian, $10^{-33}\,cm$ at $t = 10^{-43}\,s$, up to $10^{-28}\,cm$ at $t = 10^{-38}\,s$ (corresponding to the scale ~10^{-2} to $10^3\,cm$ after expansion).

The formation and evaporation of PBH leads to effective violation of the baryon conservation law. There is a trivial possibility that the energy of the vanishing baryons could be used to furnish heat and entropy to the remaining baryons. But in this case the PBHs are not specific: it is possible, at least in principle, to construct an initial singularity with metric perturbations strong enough to give the needed entropy without PBH formation and/or baryon nonconservation.

Another, nontrivial, possibility is that the observed small baryonic charge asymmetry ($(B - \bar{B})/\gamma \sim 10^{-8}$) can be obtained from a totally symmetric initial situation. Laboratory experiments have firmly established the existence of CP-violation, i.e., the absolute difference of some properties (decay branching ratios, but not masses) of particles and antiparticles. But obtaining a baryon excess from a symmetric state also requires baryon charge nonconservation. Of course, stated in that way, the hypothesis appears difficult to reconcile with the stability of ordinary matter.

Zel'dovich (1976) tried to overcome these difficulties by proposing the idea that evaporation itself can be charge-symmetric, but if unstable particles and antiparticles resulting from the evaporation have different decay properties, then the partial accretion of decay products could lead to an asymmetric universe. Indeed, the hypothesis is very speculative, since all processes are thought to occur in the earlier part of the hadron era. The last hypothesis, perhaps linked to quantization of the metric, concerns the "large numbers" $\hbar c / G m_e^2 = 10^{43}$, which are obtained when one compares particle physics with gravitation.

A. M. Polyakov (1975) and colleagues Belavin et al. (1975) developed a new type of theory deeply linked with topological objects and structures. They connected changes in field topology with intermediate states not obeying the classical equations of relativistic field theory. One such change is concerned with barrier penetration (tunneling) and certain topological singularities (monopoles, instantons). In such cases, exponentially small dimensionless numbers occur: $e^{-16\pi^2/g^2} \sim 10^{-430}$, for example. (See, for more details, chapters 5 and 14.)

In gravitational theory, no dimensionless charges exist. Still, one could obtain numbers of the type $e^{8\pi^2} = 6 \times 10^{30}$. For example, one could speculate about wormholes making space non-orientable: a righthand helicity particle falls into the wormhole, and a lefthand helicity particle emerges. The wormhole plays the role of rest-mass in Dirac spin 1/2 particle theory linking together two different helicities. In a space seeded with virtual wormholes of this type, every spin 1/2 particle would acquire rest-mass. The first estimate is $m = (\hbar c / G)^{1/2} = 10^{-5} g$ (simply by dimensional arguments), but if the wormholes are tunneling one could imagine an exponentially small dimensionless coefficient bringing the particles mass into an acceptable range. Even if the particular example of a wormhole is highly uncertain and full of difficulties (neutrino with zero rest-mass and parity violation), the general statement that new types of theories, leading to the quantization of gravity, could give exponentially small (or large) numbers seems to be proved.

The Problem of the Vacuum and the Conceptual Conflict between General Relativity Theory and Quantum Mechanics

According to Einstein and his special theory of relativity (STR), the ether does not exist at all. The electromagnetic fields are not states of a medium, but are independent realities not reducible to anything else. This conception suggests itself the more readily as electromagnetic radiation, which like ponderable matter carries momentum and energy, and, according to STR, both matter and radiation are but special forms of distributed energy. This was the point of view subsequently adopted by many theorists. Einstein's assertion is clear. According to the general theory of relativity, space-time itself is a medium.[1] "To deny the ether is to assume that empty space has no physical qualities whatever." General relativity not only restores dynamical properties to empty space but also ascribes to it energy, momentum, and angular momentum. In principle, gravitational radiation could be used as a propellant. Since gravitational waves are merely ripples on the curvature of space-time, an anti-etherist would have to describe a spaceship using this propellant as getting something for nothing—achieving acceleration simply by ejecting one hard vacuum into another. This example is not as absurd as it sounds. It is not difficult to estimate that a star undergoing asymmetric collapse may achieve a net velocity change of the order of 100 to 200 km s^{-1} by this means.

But Einstein's conception went beyond this. According to him, curvature is not the only texture that the ether (that is, space-time) possesses. It must have other, more subtle textures, which, like curvature, are best described in the language of differential geometry and differential topology. In an address delivered in 1920 at the University of Leyden, Einstein laid down the following challenge to his Leiden audience:

> As to the part the new ether is to play in the physics of the future we are not yet clear.
> We know that it determines the metrical relations in the spacetime continuum, . . .
> but we do not know whether it has an essential share in the structure of the ele-
> mentary particles. . . . It would be a great advance if we could succeed in compre-
> hending the gravitational field and the electromagnetic field together as one uni-
> fied conformation.

Einstein's attack on the unified field problem, and his failure, are well known.
The equal failure of others of high stature—Weyl, Klein, Pauli, to name but
three—led to a strong reaction among theorists, and to a turning away from
such problems for many years. Yet the dream never wholly died. Two features of
Einstein's conception remained permanently compelling: the potential richness
of a geometrically based reality, and the predictive power of theories based on local
invariance groups. These ideas have survived the explosive arrival in 1926 of the
golden age of quantum mechanics, the great era of quantum electrodynamics, the
subsequent disillusionment with quantum field theory, and the turning in de-
spair to the applied arts of dispersion theory.

Near the end of his Leiden lecture, Einstein uttered some words of caution:
"In contemplating the . . . future of theoretical physics we ought not to reject the
possibility that the facts comprised in the quantum theory may set bounds to
field theory beyond which it cannot pass." These words were spoken eight years
before the birth of quantum field theory, and Einstein therefore could not have
known that the quantum theory, far from setting bounds on field theory, would
transmute and enrich it. Yet he was right in supposing that it would introduce
severe new problems.

Let's look briefly at the ways in which the general theory of relativity and the
quantum theory of fields, taken together in a new synthesis, have thus far enriched each other. One of the most striking examples of mutual enrichment is
to be found in the impact that the ideas of general relativity have had upon the
concept of "the vacuum," and, conversely, in the reinforcement that quantum field
theory has given to the idea that the vacuum may be viewed as a textured ether.
It has been known from the earliest days of quantum electrodynamics that field
strengths, in the vacuum state, undergo random fluctuations completely analogous to the zero-point oscillations of harmonic oscillators, and that when couplings
to the electron field are taken into account these fluctuations are accompanied by
pair-creation and pair-annihilation events. The vacuum is thus in a state of constant turmoil.

From Einstein's point of view, it would have been natural to regard field fluctuations as having their seat in the ether (namely, space-time continuum) and contributing to it by qualities additional to its geometrical properties. A mathematical description of the vacuum that effectively embodied this idea was given decades ago by Schwinger (1951). In the presence of an external source, a quantized field initially in the vacuum state need not stay in that state. Schwinger showed that all physical properties of the field can be derived from a knowledge of how the probability amplitude needed for the field to remain in the vacuum state varies as the source is changed. Functional derivatives of the vacuum-to-vacuum amplitude, with respect to the source, are response functions that describe how the ether reacts to external stimuli. The ether itself thus contains a complete blueprint for the field dynamics.

The ether may be probed by means other than sources. One may vary boundary conditions and external fields. For example, the vacuum-to-vacuum matrix element of the stress tensor $T^{\mu\nu}$, of any combination of fields, including the gravitational field, is given by the functional derivative

(4.1) $$\langle \text{out, vac}| T^{\mu\nu} |\text{in, vac} \rangle = -2i(\delta / \delta g_{\mu\nu}) \langle \text{out, vac}|\text{in, vac} \rangle.$$

Here $|\text{in, vac}\rangle$ and $|\text{out, vac}\rangle$ are the initial and final vacuum state vectors, respectively, and $g_{\mu\nu}$ is an external metric field, frequently referred to as *background field*, which serves as an arbitrary zero point for the quantum fluctuations of the gravitational field, and which can be used to fix the topology of the spacetime manifold. It is assumed in (4.1) that the "vacuum" states are unambiguously (although not necessarily uniquely) defined relative to the background, and that topological transitions can be described (if indeed they occur at all) only by allowing $g_{\mu\nu}$ to move into the complex plane. It is also assumed that renormalizations have been carried out to eliminate any divergences that may arise.

The analogous equation in quantum electrodynamics is

(4.2) $$\langle \text{out, vac}| J^{\mu} |\text{in, vac} \rangle = -i(\delta / \delta A_{\mu}) \langle \text{out, vac}|\text{in, vac} \rangle,$$

where J^{μ} is the current vector, and A_{μ} is the vector potential of an external or background electromagnetic field. Expression (4.2) is generally nonvanishing whenever the background fields are nonvanishing—a phenomenon known as *vacuum polarization* (see chapter 10 for a description of this phenomenon). Similarly, expression (4.1) is generally different from zero whenever the background geometry is curved. But curvature is not the only source of gravitational "vacuum polarization." *Topology also contributes.* This means that the properties of the ether depend on the whole manifold!

This fact is so striking that it is worth examining in some detail. The phenomenon was first discovered in the Casimir effect. In the course of computing Van der Waals forces between very close molecules, Casimir (1948) found that the interaction energy could be expressed as a sum of terms involving, in addition to the molecular separation distance and the internal molecular parameters, powers of the curvature of the molecular surface. One term, however, depended on neither the curvature nor the molecular details. Its presence implied that an attractive force must exist between any two parallel flat conducting surfaces in a vacuum. The effect was soon verified in the Philips laboratories. Because the force is too tiny, great care had to be taken to ensure that the surfaces were absolutely clean, neutral, and micro-flat, so that they could be brought nearly into contact without other effects intervening. The relevant field in the Casimir effects is the electromagnetic field, and the manifold involved is the plane-parallel *slab* between the conducting surfaces. In mathematical terminology, this is an incomplete manifold. Nothing prevents one from doing physics on an incomplete manifold, provided that one is supplied with appropriate boundary conditions—perfect-conductor boundary conditions in this case. The properties of the ether between the conductors are entirely determined by the field Green's functions (response functions) appropriate to the slab.

Consider first the infinite Minkowskian vacuum—the standard vacuum of particle physicists. Thus, $T^{\mu\nu}$, the operator describing the energy, momentum, and stresses in the electromagnetic field, is formally a bilinear product of operator-valued distributions (the field operators) and hence is meaningless. It is given meaning by a subtraction process that sets the expectation value $\langle T^{\mu\nu} \rangle$ equal to zero in the Minkowskian vacuum. This subtraction corresponds to ignoring the zero-point energy of the field oscillators, but it is by no means arbitrary. In fact, $\langle T^{\mu\nu} \rangle$ *must* vanish in an empty Minkowskian spacetime if quantum field theory is to be ultimately consistent with general relativity. The Minkowskian vacuum serves as a standard by which all other vacua, whenever possible, are to be compared.

Now suppose that a single, infinite plane conductor is introduced into the Minkowskian vacuum. One may imagine it to be brought adiabatically from infinity, so that the field suffers no excitations but remains in its ground state. The manifold of interest has become an infinite half-space. Introduce Minkowski coordinates x^μ, $\mu = 0, 1, 2, 3$, oriented so that x^3-axis is perpendicular to the plane of the conductor. By considerations of symmetry, it is clear that $\langle T^{\mu\nu} \rangle$ must be diagonal and independent of x^0, x^1, and x^2. Moreover, because a perfect conductor remains a perfect conductor in any state of motion parallel to its surface, the vacuum stresses in its vicinity must look the same no matter how rapidly one is skimming over its surface. That is to say, the ether always keeps its relativistic properties, and hence $\langle T^{\mu\nu} \rangle$ must be invariant under Lorentz transformations that correspond to boosts parallel to the (x^1, x^2)-plane. This means that the first three rows and columns of $\langle T^{\mu\nu} \rangle$ must be proportional to the metric tensor of a $(2 + 1)$-dimensional Minkowskian space, namely diag $(-1, 1, 1)$. If, to this inference, one adds the observation that $T^\mu_\mu = 0$ in the case of the electromagnetic field, one concludes that $\langle T^{\mu\nu} \rangle$ has the form

$$(4.3) \qquad \langle T^{\mu\nu} \rangle = f(x^3) \times \mathrm{diag}(-1, 1, 1, -3).$$

But that is not all. The form of the function $f(x^3)$ too may be deduced. For this, one invokes the conservation law $\langle T^{\mu\nu}_{\ ,\nu} \rangle = \langle T^{\mu\nu} \rangle_{,\nu} = 0$. In particular,

$$(4.4) \qquad 0 = \langle T^{3\nu} \rangle_{,\nu} = -3f'(x^3),$$

which implies that f is a constant, independent of x^3. Now $\langle T^{\mu\nu} \rangle$ has the dimensions of energy density. The only fundamental constants that enter into the theory are $\hbar$ and c. To obtain a constant having the dimensions of energy density, one needs also a unit of length, mass, or time. No natural units with these dimensions exist in the present problem. Therefore, one can only conclude that $f = 0$, and hence $\langle T^{\mu\nu} \rangle = 0$ in an infinite half-space.

All the above arguments concerning the form of $\langle T^{\mu\nu} \rangle$ hold equally well for the slab manifold, except that there is now a natural unit of length—the separation distance, a, between the parallel conductors. In the region between the conductors, therefore, we expect

$$(4.5) \qquad \langle T^{\mu\nu} \rangle = f(a) \times \mathrm{diag}(-1, 1, 1, -3).$$

The form of the function $f(a)$ may be determined by considering the work required to separate the conductors adiabatically. From the infinite half-space analysis, one knows that the conductors experience no forces from the outside. There is an internal force, however, of amount $3f(a)$ per unit area, tending to pull them together. If the conductors are moved a distance da farther apart, an amount of work $dW = 3f(a)da$ per unit area must be supplied. This must show up as an increase in the energy per unit area. $E = -af(a)$. Setting $dW = dE$ and integrating, one immediately obtains

(4.6)
$$f(a) = A/a^4,$$

where A is some universal constant.

It will be observed that the energy density in the ether between the conductors is negative. It is a tiny energy, too small by many orders of magnitude to produce a gravitational field that anybody is going to measure. Yet one can easily construct gedankenexperiment in which the law of conservation of energy is violated, unless this energy is included in the source of the gravitational field. It turns out that the energy density in the quantum ether is often negative. The quantum theory therefore violates the hypotheses of the famous Hawking-Penrose theorems concerning the inevitability of singularities in space-time, which imply the ultimate breakdown of classical general relativity.

Chapter 5

Topology and Curvature as Sources of Vacuum Fields

In this chapter we investigate how the topological properties of space-time could influence the physical presence of field energy and particles matter in the physical vacuum. The "physical vacuum" itself could have topological properties specifically related to the role played by some geometrical non-Riemannian structures, and by global topological Riemannian features as well. It is now recognized that topological coherent structures (fields and particles, along with fluctuations) can be placed in correspondence with the concept of topological changes, in terms of processes that are cohomologically covariant when applied to certain classes of manifolds (see chapter 14 for more details on this topic).

For example, from the cosmological standpoint, although the observed universe appears to be geometrically flat, it could have a complex topology. A constant-time slice of the space-time manifold could be a torus, Möbius strip, Klein bottle, or others. This global topology of the universe imposes boundary conditions on quantum fields and affects the vacuum energy density via the Casimir effect. In a space-time with such a nontrivial topology, the vacuum energy density is shifted from its value in a simply-connected space-time. This means that the vacuum expectation value of the stress-energy tensor for a massless scalar field is calculated in several multiply-connected flat and homogeneous space-times with different global topologies. Thus, it has been found that the vacuum energy density is lowered relative to the Minkowski vacuum level in all space-times, and that the stress-energy tensor becomes position-dependent in space-times that involve reflections and rotations.

Besides general relativity and quantum field theory, as usually practiced, a third sort of idealization of the physical world has attracted a great deal of attention in the last two decades. These are called *topological quantum field theories*

(TQFTs). A TQFT is a *background-free quantum theory with no local degree of freedom*. A good example is quantum gravity in 3-dimensional space-time. First, let us recall some features of *classical* gravity in 3-dimensional space-time. Einstein's equations predict qualitatively quite different phenomena, depending on the dimension of space-time. If space-time has four or more dimensions, Einstein's equations imply that the metric has local degrees of freedom. In other words, the curvature of space-time at a given point is not completely determined by the flow of energy and momentum through that point: it is an independent variable in its own right. For example, even in the vacuum, where the energy-momentum tensor vanishes, localized ripples of curvature can propagate in the form of gravitational radiation.

The absence of local degrees of freedom makes general relativity far simpler in 3-dimensional space-time than it would be in higher dimensions. Perhaps surprisingly, this is somewhat interesting, owing to the presence of global degrees of freedom. For example, if we chop a cube, we get a space-time with a flat metric, and thus a solution of the vacuum Einstein equations. If we do the same starting with a larger cube, or a parallelepiped, we get a different space-time, one that also satisfied the vacuum Einstein equations. The two space-times are *locally* indistinguishable, since locally both look just like flat Minkowski space-time. But they can be distinguished globally, for example by measuring the volume of the whole space-time, or by studying the behavior of geodesics that wrap all the way around the torus.

Since the metric has no local degrees of freedom in 3-dimensional general relativity, this theory is much easier to quantize than is the physically relevant 4-dimensional case. In the simplest situation, where we consider "pure" gravity without matter, we obtain a background-free quantum field theory with no local degrees of freedom whatsoever, a TQFT. A TQFT describes a world quite difficult to imagine, one without local degrees of freedom. In such a world, nothing local happens, and the state of the universe can only change when the topology of space itself changes.

Here, let us consider in particular the relationship between the matter-energy operators and the global properties of some "complete" topological manifolds. In order for the energy-momentum-stresses operator $\langle T^{\mu\nu} \rangle$ to be nonvanishing in a flat empty space-time, it is not necessary that the manifold be incomplete, as is the case in the Casimir effect.

Let us consider, more specifically, the relationship between the matter-energy operators and the global properties of some "complete" topological manifolds. Complete manifolds also exhibit the phenomenon. For example, consider the manifold

$R \times \Sigma$, where the "slices," Σ, are flat, space-like Cauchy hypersurfaces having any one of the following topologies: $R^2 \times S^1$, $R \times T^2$, T^3, $T^n R \times K^2$, etc., where T^n is the n-torus, K^2 is the 2-dimensional Klein bottle, etc. The slice $\Sigma = R^2 \times S^1$ bears the closest resemblance to the Casimir example. The only difference is that, instead of imposing conductor boundary conditions on the faces of the slab, one imposes periodic boundary conditions. The operator $\langle T^{\mu\nu} \rangle$ again takes the form of (4.5), (4.6), where a is now the period of the coordinate x^3. The renormalized Green's function is easily computed, and one finds in this case that $A = \pi^2/45$. The cases $\Sigma = R \times T^2$ and $\Sigma = T^3$ are more complicated. Although $\langle T^{\mu\nu} \rangle$ is coordinate-independent, its form is no longer given by (4.5) and (4.6), but depends on the ratios of the various coordinate periodicities. In the case $\Sigma = R \times K^2$, $\langle T^{\mu\nu} \rangle$ is not even coordinate-independent, but is itself periodic (and smooth).

One advantage in studying quantum field theory on complete manifolds is that any field may be selected. One need not worry about the question: what boundary conditions are analogous to perfect-conductor boundary conditions for the electromagnetic field? Scalar and spinor fields, and even the gravitational field, may be introduced. The spinor field is of particular interest, because on some manifolds, for example on $\Sigma = R^2 \times S^1$, one may introduce spinor fields that are homotopically inequivalent. This means that more than one "vacuum" state can be defined.

Situations of this type were studied in the 1980s in connection with kinks, solitons, and instantons. General relativity, with the richness of alternative topologies that it allows, increases the variety and complexity of these situations. Moreover, as a model for other (usually simpler) field theory, it has drawn attention to a fact that had often been overlooked earlier, namely that the configuration space of any set of interacting fields is itself a Riemannian manifold or, more generally (if fermion fields are involved), a graded Riemannian manifold possessing a metric that is determined (at least in part) by the field Lagrangian. The topology of this manifold by no means must be trivial.

In all the above examples, the background manifold is flat; the energy and stresses in the ether are entirely due to topology. Not all topologies admit flat metric—for example, $\Sigma = S^3$ or $\Sigma = R \times S^2$. In these cases too, $\langle T^{\mu\nu} \rangle$ is nonvanishing. But the "vacuum polarization" is no longer exclusively produced by topology; curvature plays a role. Curvature also complicates the renormalization algorithm and gives rise to the phenomenon of trace anomalies: the formal identity $T^\mu_\mu = 0$, valid for conformally invariant classical field theories, fails for quantized fields when curvature is present. We remark here only that trace anomalies are similar in a number of respects to the axial-vector current anomalies of weak-interaction theories in which, again, a formal identity (a divergence identity) fails in the quantum theory.

Let us first add a few general remarks, and then some more specific reflections. It is well known that in field theory dealing with forces such as gravitation or electromagnetism, empty space—the vacuum—alters its nature in the presence of the relevant force. This alteration takes the form of a geometrical distortion that will change the trajectory of any material object that enters the appropriate position of space. This is what we might call the *classical theory* of the vacuum, classical in the sense of pre-quantum theory. As we saw, quantum theory requires us to modify our physical ideas in profound ways, and we must inevitably start by reexamining the classical vacuum. Quantum field theory attempts to deal with the classical force-fields in a quantum-mechanical way, and the *quantum vacuum* that emerges from this theory is a complex and mysterious structure that stretches mathematics to its utmost limits. Quantum fields fluctuate and convert themselves into particles in a bewildering manner, indicating in particular the fact that the conventional separation between force and matter cannot be maintained. Still more important (and more surprising) is the fact that the quantum vacuum does have interesting geometrical features, but these relate not to the traditional geometry of Euclid, Riemann, etc., involving measurement, but to that modern branch of the subject known as *topology*, which is concerned with qualitative properties of space. Unlike measurement, which can be conducted on a small local scale, topological features are visible only on a "global" scale.

This relation between topology and the quantum vacuum has been recognized quite recently, and its full implications have been explored in just the last three decades. It is clear that a deeper understanding of the universe rests on the possibility of further elucidating the profound connection between geometry and physics. Let us briefly describe a few examples that illustrate how topology relates to quantum theory.

THE AHARONOV-BOHM EFFECT: THE IMPORTANCE OF GLOBAL GEOMETRIC EFFECTS IN PHYSICS

A fundamental experiment, first suggested by Aharonov and Bohm forty years ago, consists in sending a beam of electrons around a solenoid carrying a magnetic flux. The experiment shows that the electrons exhibit interference patterns, depending on the strength of the magnetic flux. Thus, the electrons are physically affected by the magnetic field, even though the field lies entirely inside the solenoid and the electrons nonetheless travel in the exterior region, that is, in regions that are inaccessible to the particles. This differs from the case of classical mechanics wherein the force is zero if the magnetic field vanishes. This Aharonov-Bohm effect (1959) therefore shows that, even in a force-free region, there are physical effects. These effects are quantum mechanical, inasmuch as they correspond to phase shifts in the wave function of the electron, and they are topological in origin, since there is a cylindrical hole in the force-free region. In general, the profound consequence of the Aharonov-Bohm effect is that knowledge of the classical electromagnetic field acting *locally* on a particle is not sufficient to predict its quantum-mechanical behavior, since it depends in an essential way on the global shape and topological properties of the physical medium. The Aharonov-Bohm effect thus illustrates a first example of a not-simply-connected configuration space.

The magnetic Aharonov-Bohm effect is also closely related to Dirac's argument that the existence of a magnetic monopole necessarily implies that both electric and magnetic charges are quantized. A magnetic monopole implies a mathematical singularity in the vector potential, which can be expressed as an infinitely long Dirac string, of infinitesimal diameter, that contains the equivalent of all of the $4\pi g$ flux from a monopole "charge" g. Thus, assuming the absence of an infinite-range scattering effect by this arbitrary choice of singularity, the requirement of single-valued wave functions necessitates charge-quantization: $2qg/\hbar c$ must be an integer (in *cgs* units) for any electric charge q and magnetic charge g.

Let's now explain in more technical terms the Aharonov-Bohm effect resulting from the interference of electron beams. First, we need the action for a particle with charge q moving in a magnetic field. This is given by

$$(5.1) \quad S[\mathbf{x}(t)] = \int_{t_i}^{t_f} dt\, 1/2\, m\, (d\mathbf{x}/dt)^2 + q d\mathbf{x}/dt \cdot \mathbf{A} = S_0[\mathbf{x}(t)] + \int_{t_i}^{t_f} q d\mathbf{x}/dt \cdot \mathbf{A} dt.$$

Note that what appears in this expression is the vector potential $\mathbf{A}$ and *not* the magnetic field $\mathbf{B}$. This is crucial, because even though $\mathbf{B} = 0$ outside the solenoid, $\mathbf{A} \neq 0$ in that region.

The time integral of $\mathbf{A}$ can then be converted to a line integral,

$$(5.2) \qquad\qquad \int_{t_i}^{t_f} q d\mathbf{x}/dt \cdot \mathbf{A} dt = q \int_{\mathbf{x}_i}^{\mathbf{x}_f} q d\mathbf{x} \cdot \mathbf{A},$$

which allows us to write the action as

$$(5.3) \qquad\qquad S[\mathbf{x}(t)] = S_0 + q \int \mathbf{A} \cdot d\mathbf{x}.$$

The amplitude sufficient for the electron to travel from the initial point to the detector is thus given by

(5.4) $\mathcal{A}(i \rightarrow f) = \langle \mathbf{x}_f, t_f | \mathbf{x}_i, t_i \rangle = \int_{\text{paths}} \mathcal{D}[\mathbf{x}(t)] e^{i/\hbar S[\mathbf{x}(t)]} = \int_{\text{paths}} \mathcal{D}[\mathbf{x}(t)] e^{i/\hbar S_0 + q \int \mathbf{A} \cdot d\mathbf{x}}.$

We can divide this path integral into two pieces, one that sums up paths that travel "above" the solenoid and another that sums up paths that travel "below" the solenoid, as follows:

(5.5) $\mathcal{A}(i \rightarrow f) = \int_{\text{above}} \mathcal{D}[\mathbf{x}(t)] e^{[i/\hbar S_0 + \int \mathbf{A} \cdot d\mathbf{x}]_{\text{above}}} + \int_{\text{below}} \mathcal{D}[\mathbf{x}(t)] e^{[i/\hbar S_0 + q \int \mathbf{A} \cdot d\mathbf{x}]_{\text{below}}}.$

Each path integral has an exponential factor that depends on the magnetic field through the vector potential, $e^{iq/\hbar \int \mathbf{A} \cdot d\mathbf{x}}$. But because $\mathbf{B} = \nabla \times \mathbf{A} = 0$ in the region outside the solenoid where the paths are located, the line integral of $\mathbf{A}$ depends only on the endpoints $\mathbf{x}_i$ and $\mathbf{x}_f$ and *not* on the specific path between them. Because the exponential factor containing $\mathbf{A}$ is independent of the paths that we are integrating over, the paths can be pulled outside of the integral, yielding

$$\mathcal{A}(i \rightarrow f) = [e^{iq/\hbar \int \mathbf{A} \cdot d\mathbf{x}}]_{\text{above}} \int_{\text{above}} \mathcal{D}[\mathbf{x}(t)] e^{[i/\hbar S_0]_{\text{above}}}$$

(5.6) $$+ [e^{iq/\hbar \int \mathbf{A} \cdot d\mathbf{x}}]_{\text{below}} \int_{\text{below}} \mathcal{D}[\mathbf{x}(t)] e^{[i/\hbar S_0]_{\text{below}}}.$$

The probability for electrons to reach the detector is given by the absolute square of the amplitude. Using abbreviated notation, this is

$$P(i \rightarrow f) = |\mathcal{A}(i \rightarrow f)|^2 = \langle \mathbf{x}_f, t_f | \mathbf{x}_i, t_i \rangle|^2 = |\int_{\text{above}}|^2 + |\int_{\text{below}}|^2$$

(5.7) $$+ 2\mathcal{R} \left(e^{-iq/\hbar \int \mathbf{A} \cdot d\mathbf{x}_{\text{above}}} e^{iq/\hbar \int \mathbf{A} \cdot d\mathbf{x}_{\text{below}}} \int_{\text{above}}^* \int_{\text{below}} \right)$$

$$= |\int_{\text{above}}|^2 + |\int_{\text{below}}|^2 + 2\,Re \left(e^{iq/\hbar \int \mathbf{A} \cdot d\mathbf{x}} \int_{\text{above}}^* \int_{\text{below}} \right).$$

The last line combines a path below the solenoid with the negative of a path above, to form closed line integrals of the vector potential around the solenoid. Evaluating this situation, using vector calculus theorems, yields $\int \mathbf{A} \cdot d\mathbf{x} = \int_{\text{surface}}(\nabla \times \mathbf{A}) \cdot d\mathbf{a} = \int_{\text{surface}} \mathbf{B} \cdot d\mathbf{a} = \Phi$, where Φ is the magnetic flux passing through the closed loop, which is exactly the flux inside the solenoid. Thus the magnetic field dependence of the probability that electrons will reach the detector will have a contribution from the cross-terms that is proportional to a sine and/or cosine of $q\Phi/\hbar$. This is how the magnetic field dependence arises in our Aharonov-Bohm experiment, even though $\mathbf{B} = 0$ in all regions where the electrons travel.

In short, the probability includes a term dependent on the flux enclosed between the two different paths taken by the electrons.

The crucial point here is that the presence of a nonzero **B**-field in the solenoid means that there are paths in the plane that encircle the magnetic flux and others that do not. One can also imagine paths that encircle the flux several times. The number of times a closed path encircles the solenoid, called the *winding number*, is a topological property of the space in which the electrons are traveling. Thus the Aharonov-Bohm effect is a topological effect in quantum mechanics.

One might have a few objections at this point. First does this not prove that the magnetic vector potential **A** is physical? It does not, because the final result depends only on $\Phi = \int \mathbf{B} \cdot d\mathbf{a}$, not **A**! Because **A** appears in intermediate steps, you might want to ponder whether this result could be formulated in a way that makes no reference to the vector potential.

A second objection is that this may be all well and good for the simple, idealized two-dimensional example presented here, but in the real, three-dimensional world, some magnetic field may well be leaking out of the solenoid, which isn't really infinite anyway. Are the electron beams *really* probing the topology of the space? The answers emerged from an important, definitive experiment conducted by the Japanese physicist Akira Tonomura and collaborators in 1985 [Tonomura et al. (1986)]. They fabricated a toroidal ferromagnet, covered it with a superconducting layer to prevent the magnetic field from leaking out, and covered that with a copper conducting layer to prevent electron wave functions from leaking in. Then they examined the interference pattern produced by illuminating the torus with an electron beam. The part of the electron beam that passes outside the torus interferes with a reference beam to produce the horizontal interference fringes. The part of the electron beam that passes through the hole in the torus also interferes with the reference beam, but the fringes there are obviously shifted, indicating a different phase ($e^{iq/\hbar \int \mathbf{A} \cdot d\mathbf{x}}$) for the electron beam passing on the outside of the magnetic flux. In short, the electron beam passing through the hole in the doughnut acquires an additional phase, resulting in fringes that are shifted with respect to those outside the torus, thus demonstrating the observable Aharonov-Bohm effect due to the existence of vector potentials.

What is there about this experimental arrangement with the torus that causes it to reflect the situation we had previously analyzed? The toroidal configuration also has the property that the closed loops surround the magnetic field an integer number of times. In mathematical language, this means that the plane with a solenoid removed ($\mathbb{R}^2 - 0$) and three-space with a toroid removed ($\mathbb{R}^3 - T^2$) has the same first homotopy group, π_1. In the following section, we will explain the mathematical interpretation of the Aharonov-Bohm effect.

In the terms of modern differential geometry, the Aharonov-Bohm effect can be understood to be the monodromy of a flat complex line bundle. The $U(1)$-connection on this line bundle is given by the electromagnetic four-potential **A** as $\nabla = \mathbf{d} + ie\mathbf{A}$, where **A** means partial derivation in the Minkowskian space $\mathbb{M}^4$. The curvature form of the connection, $\mathbf{F} = d\mathbf{A}$, is the electromagnetic field strength, where **A** is the one-form corresponding to the four-potential. The holonomy of the connection, $e^{i\int_\gamma \mathbf{A}}$ around a closed loop γ is, as a consequence of Stokes' theorem, determined by the magnetic flux through a surface bounded by the loop. This description is general and holds inside the conductor as well as outside. Outside of the conducting tube, which is for example a longitudinally magnetized infinite metallic thread, the field strength is $\mathbf{F} = 0$; in other words, outside the thread the connection is flat, and the holonomy of a loop contained in the field-free region depends only on the winding number around the tube, and is, by definition, the monodromy of the flat connection.

In any simple connected region outside of the tube we can find a gauge transformation (acting on wave functions and connections) that gauges away the vector potential. But if the monodromy is nontrivial, there is no such gauge transformation for the whole outside region. If we want to ignore the physics inside the conductor and describe only the physics outside, it becomes natural to describe the quantum electron mathematically by a section in a complex line bundle with an "external" connection ∇ rather than an external electromagnetic field **F**. (By incorporating local gauge transformations, we have already acknowledged that quantum mechanics defines the notion of a locally flat wave-function—zero momentum density—but not that of unit wave-function.) The Schrödinger equation generalizes readily to this situation. In fact, the Aharonov-Bohm effect can operate in two simply connected regions, with cuts that pass through the tube toward or away from the detection screen. In each of these regions, we must solve the ordinary free Schrödinger equations, but in passing from one region to the other, in only one of the two connected components of the intersection (effectively in only one of the slits), we pick up a monodromy factor $e^{i\alpha}$, which results in a shift in the interference pattern.

There are many examples of path integrals for spaces with topological constraints, such as a particle in a box, on a half-line, or generally in a half-space. Here we look at *a particle on a circle*. On a circle, the coordinate is φ, $0 \leq \varphi \leq 2\pi$, with $\varphi = 0$ and $\varphi = 2\pi$ identified. A path in this system is a continuous function

$\varphi(t)$ with the identification given above. We can divide the set of all paths into subsets of paths with the same winding number. Taking $h = 1$, we can therefore write the kernel

$$(5.8) \qquad K(\varphi_2, t_2; \varphi_1, t_1) = \sum_{\varphi(t_1) = \varphi_1} e^{iS[\varphi(t)]} = \sum_{n=-\infty}^{\infty} \sum_{\varphi(t) \in g_n} e^{iS[\varphi(t)]},$$

where

$$g_n = \{\varphi(t) \mid t_1 \leq t \leq t_2, \varphi(t_1) = \varphi_1, \varphi(t_2) = \varphi_2,$$

and where $\varphi(t)$ is continuous and has winding number n.

We now assume that each term in the sum (5.8) individually satisfies the Schrödinger equation

$$(5.9) \qquad\qquad i\hbar(\partial\Psi/\partial t) = H\Psi.$$

We see that in this case that

$$(5.10) \qquad\qquad \sum_n A_n \sum_{\varphi \in g_n} e^{iS[\varphi(t)]}, A_n \in \mathbf{C}$$

can be a solution for the kernel as well. From periodicity, follows the condition $A_{n+1} = e^{i\delta} A_n, \delta \in \mathbf{R}$. With $A_0 = 1$, we get $A_n = e^{in\delta}$ and can then write

$$(5.11) \qquad K(\varphi_2, t_2; \varphi_1, t_1) = \sum_{n=-\infty}^{\infty} A_n K_n(\varphi_2, t_2; \varphi_1, t_1),$$

where $K_n(\varphi_2, t_2; \varphi_1, t_1)$ is the kernel for all paths with n loops. For each K_n, we can carry the Lagrangian (we can do that, since we can define a smooth mapping from S^1 to $\mathbf{R}$) and, for a free particle, use

$$(5.12) \qquad K(b, a) = [2\pi i\hbar(t_b - t_a)/m]^{-1/2} \exp{(im(x_b - x_a)^2/2\hbar(t_b - t_a))}.$$

Inserting (5.12) in (5.11), we obtain the kernel for a free particle moving on a circle:

$$(5.13) \qquad \begin{aligned} K(\varphi_2, t_2; \varphi_1, t_1) &= \sum_{n=-\infty}^{\infty} (I/2\pi i(t_2 - t_1))^{1/2} \\ &\exp[in\delta + (iI/2(t_2 - t_1))((\varphi_2 - \varphi_1) - 2n\pi))^2]. \end{aligned}$$

VACUUM POLARIZATION AND THE TOPOLOGY
OF THE QUANTUM VACUUM

Topological phenomena are of great interest and importance because of their universal nature and their connections with general properties of space-time, on the one hand, and their numerous practical aspects, on the other hand. Since the discovery of Bohm and Aharonov forty years ago (see the preceding section), it has become clear that topology has to do with the fundamental principles of quantum theory. At present, much attention is accorded the study of nonperturbative effects in quantum systems, arising as a consequence of the interaction of quantized fields with a topological nontrivial classical field background.

The dependence of vacuum polarization effects on the geometry and topology of the base space was discovered primarily by Yuri A. Sitenko in 1997. In particular, it was shown that there exists a field-theoretical analogue of the Bohm-Aharonov effect: singular configurations of the external magnetic field strength induce vacuum charge on noncompact topologically nontrivial surfaces, even in cases when the magnetic flux through such surfaces vanishes. This is due to the fact that in some noncompact, essentially curved or topologically nontrivial spaces the asymptotic of the axial-vector current becomes nontrivial and contributes to the induced vacuum charge. As a result, the latter depends on global geometric characteristics of space, as well as on global characteristics of external field strength (total anomaly). A similar result was later obtained for the induced vacuum angular momentum. Thus, singular configurations of the magnetic field strength are shown to induce vacuum charge and angular momentum on noncompact topological nontrivial surfaces, as well as in cases when the magnetic flux through such surfaces vanishes.

Recently, a comprehensive and self-consistent study of nonperturbative vacuum polarization in the background of nontrivial topology was completed [see Sitenko (1997) and Sitenko and Rakitsany (1998)]. The method of self-adjoint extensions was employed to determine the most general conditions on the boundary between the spatial regions that are accessible and inaccessible for the quantized matter fields. In view of the wide use of such concepts as strings and p-branes in modern physics, the background of a singular magnetic vortex (string) is of special interest. All vacuum polarization effects in this background are determined. The nonvanishing vacuum characteristics are shown to be charge, current, energy, spin, and angular momentum. It has been found [Sitenko (2000)] that local and global vacuum characteristics are dependent on the magnetic vortex flux and the self-adjoint extension parameter. These results yield a completely new realization of the Bohm-Aharonov effect in quantum field theory.

THE FIELDS AND GEOMETRY OF BUNDLES
AND CONNECTIONS

The geometric concepts of bundles and connections are at the core of the most fundamental physical theories. As we know, gravitation in general relativity can be formulated in terms of bundles and connections. There is also a quantum-mechanical phenomenon that requires bundles, namely the Dirac magnetic monopole (which, however, has never been shown to exist physically). Consider then an electron moving in $\mathbb{R}^3 - \{O\}$ in the field of a magnetic monopole of strength q fixed at the origin. The **B** field for this monopole is $\mathbf{B} = (q/r^2)\, \partial/\partial r$, that is,

$$(5.14) \qquad \mathcal{B}^2 = i_\mathbf{B}\mathrm{vol}^3 = \mathbf{d}[q(1-\cos\theta)d\phi].$$

Thus

$$(5.15) \qquad \mathcal{A}_U^1 = q(1 - \cos\theta)d\phi$$

in the region $U = \mathbb{R}^3 - \{\text{negative } z \text{ axis}\}$. We shall need also to consider points on the negative z axis (except for the origin). In the region $V = \mathbb{R}^3 - \{\text{positive } z \text{ axis}\}$, we can use $\theta' = \pi - \theta$ and $\phi' = -\phi$ as coordinates and obtain

$$(5.16) \qquad \mathcal{A}_V^1 = q(1 - \cos\theta)d\phi.$$

Maxwell's equations hold everywhere on $\mathbb{R}^3 - \{0\}$. Since $\mathcal{A}_U^1$ does not agree with $\mathcal{A}_V^1$ in $U \cap V = \mathbb{R}^3 - \{z \text{ axis}\}$, we shall be forced to introduce the electromagnetic bundle and connection. Let the transition function for this *monopole bundle* be

$$(5.17) \qquad c_{VU} = \exp(-2ieq\phi/h).$$

Note that this is not single-valued unless Dirac's *quantization condition* that

$$(5.18) \qquad 2eq/\hbar \text{ must be an integer}$$

is satisfied. Since there are only two patches U and V, equation

(5.19)
$$\psi_V = c_{VU}\,\psi_U$$

is automatically satisfied. If (5.17) holds, the monopole bundle will exist.

That c_{VU} is not in general single-valued is a reflection of the fact that in this case $U \cap V = \mathbb{R}^3 - \{\text{entire } z \text{ axis}\}$ is certainly not simply-connected (more to the point, its first Betti number does not vanish). It is true that by using more sets (whose intersections are simply connected) to cover $\mathbb{R}^3 - \{0\}$, we could find transition functions that would be single-valued without requiring (5.18), but it would turn out that it is not possible to satisfy the crucial equation

(5.20)
$$c_{UV}c_{VW}c_{WU} = 1.$$

In fact, one can prove, from a general Gauss-Bonnet theorem, that for *any* complex line bundle over $\mathbb{R}^3$-origin, the curvature *must* satisfy

(5.21)
$$i/2\pi \int_{S^2} \theta^2 = \text{integer.}$$

The unit sphere S^2 is a generator for the second homology group $H_2(\mathbb{R}^3 - \{0\}, \mathbb{Z})$. It is first possible to prove that

(5.22)
$$i/2\pi \int_{M^2} \theta^2 = \text{integer}$$

in the geometrical case when the complex line bundle is the tangent bundle to the oriented closed surface M^2. For the monopole bundle, it follows from the fact that the *curvature* of the connection is essentially the electromagnetic field 2-form,

(5.23)
$$\theta = d\omega = -\,(ie/h)\,dA^1 = -\,(ie/h)\,F^2 = -\,(ie/h)\,[\mathcal{E}\wedge dt + \mathcal{B}],$$

that we have

(5.24)
$$\theta = (ie/h)\mathcal{B} = -\,(ie/h)i_{\mathbf{B}}\,\text{vol}^3 = -\,(ieq/r^2 h)i_{\partial/\partial r}\,\text{vol}^3,$$

and the integral in (5.21) becomes

(5.25)
$$-(i/2\pi)\,(ieq/h)\,(4\pi) = (2eq/h).$$

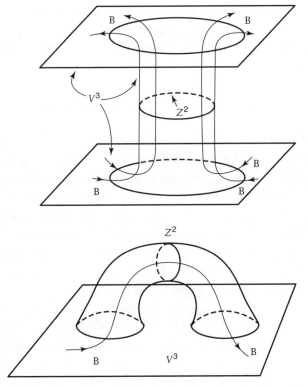

Fig. 4. I have indicated here a V^3 that consists of two separated horizontal sheets (two "separate" universes) that are joined by a wormhole cylinder $S^2 \times [0, 1]$; I have also indicated one of the spherical sections Z^2 encircling the "throat" of the wormhole. A magnetic field goes from the bottom sheet through the lower "mouth," threads its way through the throat, and emerges from the upper mouth. In this example, $\int_Z \mathcal{B}^2 \neq 0$. The drawing below is similar except that the wormhole joins two distant portions of the "same" universe, and again **B** has nonzero flux through the throat. In both cases, there is no global **A**, and the flux of **B** through the throat must be quantized in terms of e.

This yields a quantization condition, relating the charge on a monopole to that of the electron. More generally, it is possible to show that the flux $e\mathcal{B}/2\pi h$ through any closed surface, for any magnetic field, must be an integer.

I want to emphasize one final point. If there is a monopole, then from the form (5.17) we see that $\psi_V = \exp(-2ieq\phi/h)\ \psi_U \neq \psi_U$. Thus the electron wave function ψ cannot be defined (and single-valued) everywhere, and it must, rather, be considered as a *section* of the monopole bundle with at least two patches. Presumably, then, we should expect that other types of fields that interact with elementary particles might demand that wave functions be replaced by sections of bundles, just as we do not expect that every manifold should be covered by a single coordinate patch. See fig. 4.

INSTANTONS AND THE VACUUM

In Yang-Mills we may consider the *vacuum state* in which the field strength F or θ vanishes. One must not conclude that nothing of interest can be associated with such a vacuum. In the geometric analogue, we may consider a flat surface; the connection ω replaces the gauge field A, and the curvature $\theta = 0$ replaces field strength $F = 0$. In the case of the frustum of a flat cone, tangent to a 2-sphere along a small circle S^1, we may delete the spherical cap completely. This corresponds to the exterior region in the Aharonov-Bohm effect. The parallel translation about S^1 does not return a vector to itself, in spite of the fact that the connection is flat. (There is more information in the flat connection than is read off from the zero curvature alone!) Likewise, there is more information in a gauge field A for a vacuum than can be read from the vanishing field strength.

Before considering the Yang-Mills vacuum, we shall look at another geometric analogue. Suppose we have drawn the 2-dimensional analogue, a flat surface, but instead of using the "flat" (covariant constant) frame (pointing, for example, constantly in the t direction), we use a frame that is time-independent, is flat at spatial infinity, and rotates (in this case) once about the flat frame along each spatial slice. We have *gauge transformed* the flat frame e to a new one, e.g., where $g : \mathbb{R} \to U(1) = S^1$ maps each spatial slice so that $g(-\infty) = g(\infty) = 1$. (The field, i.e., connection, is again "pure gauge," $\omega = g^{-1}dg$.) We assume, again, for simplicity, that for each spatial section $t = $ constant we have $g(x) = 1$ for $|x| \geq a$ for some a. This vacuum solution in $\mathbb{R}^2$ is not deformable, while remaining flat at spatial infinity, into the identically flat frame vacuum, for the following reason.

The function g maps the spatial slice $\mathbb{R}$ into S^1. We may stereographically project $\mathbb{R}$ onto a circle S^1–(north pole) by projecting from the north pole. In this way, we may consider g as being defined on S^1–(north pole). Since g is identically 1 in some neighborhood of the pole, we can extend g to the entire circle S^1. (This can be thought of in the following way. By identifying all x for $|x| \geq a$ with the point $x = a$ on the section $t = $ constant, this section becomes topologically a circle S^1. We have "compactified" $\mathbb{R}$ to a circle, and since $g = 1$ for $|x| \geq a$, g extends to this compactification.) This gives, for each t, a map $g : S^1 \to U(1) = S^1$, which in this case has degree 1 by construction. If our vacuum solution were to be deformable to a flat vacuum solution, while keeping $|x| \geq a$ flat, then $1 = \deg g :$ $S^1 \to U(1) = S^1$ would have to equal that of the flat vacuum case, which clearly has degree 0. This is a contradiction. We thus have two inequivalent vacua. Similarly, we could obtain a vacuum frame that winds k times around the flat frame.

In the 4-dimensional Yang-Mills case (with $G = SU(2)$), there will likewise be an infinity of inequivalent vacua, each one characterized by the degree or "winding number" of the map $g : S^3 \to SU(2) = S^3$ arising from the spatial slice $\mathbb{R}^3$ "compactified" to S^3. Physicists then interpret an instanton with winding number k, that is, degree k given in the above map (and this map does indeed have degree, called the *winding number of the instanton*), as representing a nonvacuum field tunneling between a vacuum at $t = -\infty$, with winding number n, and a vacuum at $t = +\infty$ with winding number $n + k$.

We have seen why $g : S^3 \to SU(2)$ has a degree. To understand why $g : S^3 \to SU(n)$, $n \geq 2$, has an associated "degree," and to understand other global results (Chern's results) when there are topological singularities, we would need to delve more into topology, in particular the topology of Lie groups, "homotopy groups," and "characteristic classes." However, we shall not address these topics in the present work.

TOPOLOGY AND QUANTUM FIELD THEORY: AN OVERVIEW

Beyond the single cylindral hole or flux tube of the Aharonov-Bohm experiment, we can consider a complicated knotted configuration of tubes. The study and classification of such knots is a typical and difficult problem in topology. The more elaborate the knot, the more intricate is the structure of the external vacuum. In fact, over the past twenty years we have come to see that the topology of knots is intimately related to physics, and that the formal apparatus of quantum field theory can be used to solve difficult topological problems in the theory of knots. These developments strongly suggest that topological aspects of 3-dimensional space, as manifested by knots, should play some fundamental role in quantum physics. Moreover, this role should involve the basic physical framework, prior to the introduction of matter or force, in other words "the vacuum." The relation between the topology of knots and quantum field theory was elaborated in the late 1980s [Witten (1989)], and the different theories may be described generically as "topological quantum field theories." The essential characteristic of such theories is that they have the formal structure of a quantum theory (e.g., dealing with discrete spaces and probabilities), but the information they produce is purely topological (e.g., information about the nature of a knot). There is no measurement or genuine dynamics in such theories; they deal with nature at a more primordial level. At a later stage, one may imagine superimposing a more conventional physical theory onto the topological background.

The point of departure in Witten's work was the beautiful results by Vaughan Jones (1987), who found many new invariants of knots. Essentially, Jones first draws the knot in the plane as the closure of a braid. To compute the invariants, he applies a complicated formula that depends on the precise form of the braid. The result is an invariant for knots. The key point is that a knot is the closure of many different braids. Jones's formula depends on choosing an arbitrary braid to represent the knot. Thus the formulae are really formulae for braids, which happen to yield the same answer on all braids representing the same knot. It therefore becomes difficult to see which properties of the knot itself are actually expressed by these formulae.

For several years, finding an intrinsic, geometric definition of the Jones polynomials, without referring to braids, was an unsolved problem. This problem has been resolved by Witten, who describes a quantum field theory for a 3-manifold embracing knots. The theory has no unnecessary ingredients; only the ambient 3-manifold and an oriented collection of knots are given. Witten then produces a great many invariants, all resulting from a path integral over fields. Conformal field theory allows for a careful analysis of the properties of such integrals, and Witten shows that the invariants should coincide with the Jones polynomials. Because Witten's theory is intrinsic, it lends itself to generalizations. Probably the most profound one is that he can define invariants for knots in arbitrary 3-manifolds, where the concept of braids is not in play.

Another (actually the first) topological quantum field theory to have been discovered was not the one related to knots in 3-dimensional spaces, but a more intricate one related to the deep mathematical results of Simon Donaldson (1983) on 4-dimensional geometry, whose physical significance has then been elucidated by Witten. The Donaldson-Witten theory entails vacuum solutions. To begin our discussion of 4-dimensional space, pick a 4-dimensional manifold X bearing a metric. Instead of looking at Maxwell fields, we look for gauge fields on X that satisfy the Yang-Mills equations. The Yang-Mills equations are analogous to the Maxwell equations in cases where one studies gauge fields describing the weak or strong interaction. Such fields on X are called *vacuum solutions*, much in the spirit of the classical vacuum discussed above. A vacuum solution has a quantized charge $k = 0, 1, \ldots n$. If we fix the charge, we find a complicated space of solutions called the Yang-Mills moduli space. Because this space depends on X and the metric on X, it is not yet a topological invariant of X. The moduli space presents all kinds of loops and holes. Some holes in this space will disappear if we vary the metric on X, while others persist. Clearly, the holes that persist must be (differential) topological invariant of X.

Donaldson's remarkable observation (1983) was that such loops do exist, and that they contain deep information about the 4-manifold. Donaldson's theory is this one using classical vacuum to say something about pure space. It is a very powerful theory. It shows, for example, that even on good old 4-dimensional Euclidean space there are infinitely many inequivalent ways of defining differentiability of functions; that is, the differential topology of 4-dimensional vector spaces is highly nontrivial.

Witten has developed a complicated supersymmetric quantum field theory, such that the expectation values of certain observables give the relevant information about persisting loops. Witten's quantum field theory in its original version depends on a metric on X, just as Donaldson's theory does. He shows that his expectation values do not change if we vary the metric, thereby demonstrating that they are topological invariants—that is, invariants of pure space. But his theory is full of all kinds of fields and superfields. So again, a nonvacuous theory gives information about the ultimate vacuum, pure space.

In gauge theory the issue is to find anti-self-dual connections on some bundle over the 4-manifold X. These are precisely the connections where the self-dual part of the curvature vanishes. Thus we are looking for zeroes of a function s defined on the space A of connections; the function assigns to a connection the self-dual part of its curvature, which is an element of an infinite-dimensional vector space V. Gauge invariance enters the picture in the following way. If we apply a gauge transformation to the connection, then the self-dual part of the curvature is conjugated by this gauge transformation. Thus the function s does not descend to a function on the quotient space A/G of connections modulo gauge transformations, since its values get twisted in V. Instead, the function becomes a section of a vector bundle W over A/G. Looking for anti-self-dual connections modulo gauge transformations amounts to looking for the zeroes of a section of this bundle.

We shall now assume that there are only finitely many zeroes in these sections. In practice, this situation is not always met, but it does happen in interesting situations. Donaldson's invariant is now simply the number of zeroes, that is, the number of anti-self-dual connections, counted with signs. As indicated below (fig. 5), the nontriviality of a bundle can force a section to have zeroes. Generally, the number of zeroes of a section of a bundle E over a compact manifold M is an invariant

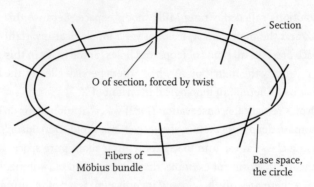

Section

O of section, forced by twist

Fibers of —
Möbius bundle

Base space,
the circle

Fig. 5. Möbius bundle with a section.

of E, called the *Euler number*. In particular, it is independent of that section. The Euler number can be computed in various ways, and this is precisely the distinction between Donaldson's theory and the Witten-Donaldson theory.

First of all, one can simply compute the number of zeroes in a section; in the case at hand, this amounts to solving the self-dual Yang-Mills equations. A more sophisticated approach is to integrate a differential form over the total space of E. The form is the square of the Thom class of the bundle. In the Yang-Mills case, we are integrating the square of the Thom class over the infinite-dimensional bundle W over an infinite-dimensional space A/G. Such integrals are normally called *Feynman path integrals*. This is more or less what Witten does to compute Donaldson's polynomial invariants.

I should like to make one last remark, relating to recent works on the connection between knot topology and quantum chromodynamics vacuum. First, I remind the reader that the non-Abelian gauge theory has been well known to have a nontrivial topology. In particular, the theory offers infinitely many topologically distinct vacua that can have been connected by vacuum tunneling through the instantons. The existence of both topologically distinct vacua and vacuum tunneling has played a very important role in quantum chromodynamics (QCD).

In a totally independent development, the Skyrme theory has been shown to admit a topologically stable knot that can be interpreted as a twisted magnetic vortex ring made of helical baby skyrmion (topological solitons in a $(2 + 1)$-dimensional field theory). Very interestingly, this knot has been shown to describe the topologically distinct QCD vacua. This finding is puzzling, because the knot is a physical object that carries a nonvanishing energy. It may appear strange that the knot can be related to a QCD vacuum, but the Skyrme theory itself is closely related to QCD, and both the knot and the QCD vacuum are described by the same topology $\pi_3(S^3)$ [Skyrme (1961)]. In this circumstance, one needs to know in exactly what sense the QCD vacuum can be identified as the knot. Since there exists one knot solution for each topological quantum number (up to the trivial space-time translation and the global $SU(3)$ rotation), one might suspect that the knot equation could be viewed as a gauge condition for the topologically equivalent vacua.

In fact, it has been suggested that the knot equation can be viewed as a nonlocal gauge condition that describes the maximal Abelian gauge in $SU(2)$ QCD [van Baal and Wipf (2001)]. More recently, it has been shown that the knot equation is nothing but a generalized Lorentz gauge condition that selects one representative vacuum for each class of topologically equivalent QCD vacua.

This conclusion allows for interpreting the knot as a complex vector field that couples to an Abelian gauge field, and the knot equation as an Abelian gauge condition for the complex vector field [Cho (2007)]. Cho obtains a most general expression of the vacuum and writes the knot equation completely in terms of the vacuum potential. With this in hand, he proves that the knot equation is nothing but a generalized Lorentz gauge condition of the QCD vacuum. From this, he shows that the knot equation can be viewed as an Abelian gauge for a complex vector field. Moreover, he shows that this complex vector field is uniquely determined by the Abelian gauge potential. This then allows a new interpretation of the knot, the knot as a complex vector field or an Abelian gauge potential. This in turn tells us that one can classify the topologically different QCD vacua by an Abelian Chern-Simon index. The knot quantum number is given by the Abelian Chern-Simon index of the magnetic potential C_μ,

$$(5.26) \qquad Q = g^2/32\pi^2 \int \varepsilon_{ijk} C_i H_{ik} d^3x,$$

which describes the nontrivial topology of the Hopf mapping $\pi_3(S^2)$ defined by $\hat{n}$. (In the $SU(2)$ QCD, $\hat{n}_i$ ($i = 1, 2, 3$) is a righthanded local orthonormal frame.)

The pre-image of the mapping from the compactified space S^3 to the target space S^2, defined by $\hat{n}$ forms a closed circle, and two pre-images of the mapping are linked together when $\pi_3(S^2)$ is nontrivial. This linking number is given by the Abelian-Chern-Simon index. It is possible to show that the same description applies to the QCD vacuum. In particular, this means that with the same index the QCD vacuum can also be classified by an Abelian gauge potential. Conversely, with the vacuum equation

$$D_\mu \omega_\nu - D_\nu \omega_\mu = 0,$$

(5.27)

$$H_{\mu\nu} = \partial_\mu C_\nu - \partial_\nu C_\mu = ig(\omega_\mu * \omega_\nu - \omega_\nu * \omega_\mu)$$

(where D_μ is the vacuum potential and ω_μ is a complex vector field), one can transform the knot quantum number above to a non-Abelian form

(5.28)
$$Q = g^2/32\pi^2 \int \varepsilon_{ijk} C_i H_{ik} d^3x,$$
$$= -g^2/32\pi^2 \int \varepsilon_{abc} C_i^a C_j^b C_k^c d^3x,$$

which proves that the knot quantum number can also be expressed by a non-Abelian Chern-Simon index. More significantly, this tells us that the Abelian Chern-Simon index is actually identical to the non-Abelian Chern-Simon index. The two have been thought to be two different things, but the previous observations tell us that they are one and the same thing and can be transformed to each other through the vacuum condition above.

ABELIAN GAUGE THEORIES AND TOPOLOGY

Some striking topological features in non-Abelian gauge theories are related to the notion of Abelian projection and the Hopf invariant. Abelian projection was introduced by 't Hooft (1981) in an attempt, through a suitable choice of gauge, to decompose a non-Abelian gauge field in its neutral and charged components. In its simplest form it involves choosing an observable $X(x)$ that transforms under gauge transformations as $g^\dagger(x)X(x)g(x)$, which can be used to diagonalize $X(x)$. This can be done smoothly when none of the eigenvalues coincide. The remaining gauge freedom is $U(1)^r$, where r is the rank of the gauge group. These gauge transformations are associated with the r neutral gauge bosons in this gauge. Singularities occur when two (or more) eigenvalues coincide, and these can be shown in three dimensions, to give rise to (generically) pointlike singularities representing magnetic monopoles, as defined with respect to the remnant Abelian gauge group.

A smoother, but nonlocal, Abelian gauge fixing can be introduced by taking an Abelian field as a background [e.g., for $SU(2)$ gauge theory the component proportional to $\tau_3 = \mathrm{diag}(1, -1)$], and imposing the background gauge condition on the charged component of the gauge field. This can be formulated by minimizing $\int |A_\mu^{ch}(x)|^2$ along the gauge orbit.

Inspired by the Abelian projection, Faddeev and Niemi (1997) attempted to identify the field $\vec{n}(x)$ (here of unit length, $|\vec{n}(x)| = 1$) in an $O(3)$ nonlinear sigma model with the local color direction for $SU(2)$ gauge theory. The hope was that the static *knotted* solutions, constructed numerically in the model originally introduced by Fadeev, were possibly related to glueballs. A difficulty is that one would not expect dynamics for a quantity that is associated with color direction, owing to gauge invariance. Furthermore, the $O(3)$ symmetry is, in general, spontaneously broken for nonlinear sigma models in $3 + 1$ dimensions. The associated Goldstone bosons are unwanted in non-Abelian gauge theories lacking matter. Nevertheless, the $\vec{n}$ field *can* be identified with an $SU(2)$ gauge field, albeit with zero field strength. Under this identification, we have $(\partial_\mu \vec{n}(x))^2 = 4|A_\mu^{ch}(x)|^2$, which hints that minimizing the $O(3)$ energy functional can be interpreted in terms of maximal Abelian gauge fixing.

The static knotted solutions, as maps $\vec{n}(\vec{x})$ from S^3 (compactified R^3) to S^2, are classified by the Hopf invariant. The pre-image of a generic value of $\vec{n}$ traces out a loop in R^3 (i.e. the collection of points $(\vec{x})$, where $\vec{n}(\vec{x}) = \vec{n}$). The linking number of any two such loops is equal to this Hopf invariant. This invariant also coincides with the winding number of the gauge function $g(\vec{x})$, such that $n^a(\vec{x})\tau_a = g(\vec{x})\tau_3 g^\dagger(\vec{x})$, where τ_a is the Pauli matrices. The associated $SU(2)$ gauge field with zero field strength is $A(\vec{x}) = g^\dagger(\vec{x})dg(\vec{x})$.

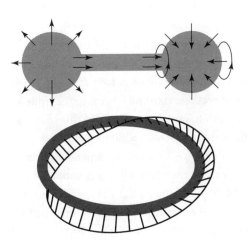

Fig. 6. A topological charge constructed from oppositely charged monopoles by rotating one of them. For a closed monopole line, the embedding of the unbroken subgroup makes a full rotation.

Such a relation between Hopf invariant and topological charge can also occur for monopoles [see Nash and Sen (1983)]. A basic monopole is characterized by a hedgehog, with the Higgs field defining the color direction, pointing outward radially. If we gauge rotate the monopole (for the spherical symmetric case, this is quivalent to a real rotation) while moving along its worldline, a particular point of the hedgehog will trace out a loop whose linking with the monopole worldline is an invariant. This approach was used by C. H. Taubes to make, from monopole fields, configurations with nonzero topological charge (Taubes, 1990). In his formulation, a monopole-antimonopole pair is created, and separated to a finite distance. Kept at this separation, one of them is rotated around the axis connecting the two. Once this is completed, they are brought together and made to annihilate themselves. The time is considered to be Euclidean, and the rotation introduces a "twist" in the field that prevents the 4-dimensional configuration from decaying to $A = 0$. The obstruction is precisely the topological charge, whose value is equal to the number of rotations.

This process can just as well be described (fig. 6) in terms of a closed monopole loop (in the same way that the Wilson loop is associated with the creation, propagation, and annihilation of a heavy quark-antiquark pair). The identification with the Hopf invariant should be understood in the following sense: the orientation of the monopole is described by $SU(2)/U(1) \sim S^2$ at each point on the monopole loop (S^1), describing a twisted S^2 bundle over S^1, e.g., making the total space into an S^3 for one full "frame" rotation. This is the Hopf fibration, although we have interchanged fiber and base space as compared with the usual formulation.

At finite temperature, A_0 plays the role of the Higgs field, and the calorons (periodic instantons) provide an explicit example of this Taubes winding, when it has a nontrivial value of the Polyakov loop at infinity. This Polyakov loop is independent of the directions at infinity, because the finite action of the caloron forces the field strength to vanish at infinity. In this case, the caloron splits into [n for $SU(n)$] constituent monopoles, which are the basic spherically symmetric BPS monopoles. Of these, $n - 1$ are time-independent, whereas the time dependence of the others exactly coincides with the rotation (at a uniform rate) of the Taubes winding. In terms of the Abelian projection and the introduction of a composite Higgs field, the relation of the Hopf invariant to twisted monopole loops has been extensively studied by O. Jahn (2000).

Chapter 6

The Dirac "Full-of-Particles Sea" Idea and the Vacuum in Quantum Field Theory

The tendency of the wave function in nonrelativistic quantum mechanics to produce so many interpretations has been one of the most important theoretical problems of quantum field theory (QFT). Certain difficulties demonstrated that the classical field was not an appropriate candidate for the ontology of nonrelativistic quantum mechanics. Let us first ask: What is the ontology of quantum theory? In a very schematic way, one can say that De Broglie and Schrödinger held a realistic interpretation of the wave function; they assumed a field ontology and rejected the particle ontology because, they argued, quantum particles obeying quantum statistics showed no identity, and thus were not classically observable individuals. In his probabilistic interpretation, Max Born rejected the reality of the wave function, deprived it of energy and momentum, and assumed a particle ontology. Heisenberg interpreted the wave function as a potential.

The introduction of fermion field quantization by Jordan and Wigner (1928) initiated further radical changes in the interpretation of QFT. First, a realistic interpretation replaced the probabilistic interpretation of the wave function: in their formulation, the wave function had to be interpreted as a kind of substantial field. Otherwise, the particles, seen as the quanta of the field, could not derive their substantiality from the field, leaving unsolved some of the original difficulties faced by Schrödinger's realistic interpretation.

Second, the field ontology replaced the particle ontology: the material particles (fermions) were no longer regarded as having an eternally independent existence, but rather as being in transient excitation of the field, a quantum of the field, thus justifying the claim that QFT initiated a major variant of the field program, namely, the quantum field program.

But the reversion of this view of the real field evidently leaves a gap in the logic of models of reality and destroys the traditional conception of substance. A new conception of ontology is thus introduced. This new ontology cannot be reduced to that of classical particle ontology, because the field quanta lack both permanent existence and individuality. It also cannot be reduced to that of the classical field ontology, because the quantized field has lost its continuous existence. Twentieth-century field theories thus seem to suggest that the quantized field, together with some nonlinear fields, constitute a new kind of ontology (say, ephemeral ontology).

The new ontology of QFT was embodied in the Dirac vacuum. As an ontological background underlying the conceptual schema of quantum excitations and renormalization, the Dirac vacuum had become crucial for calculations, such as Weisskopf's calculation (1939) of the electron's self-energy and Dancoff's discussion (1939) on the relativistic corrections to scatterings. The fluctuations existing in the Dirac vacuum strongly indicate that the vacuum must be something substantial, rather than empty. But according to the special theory of relativity, the vacuum must be a Lorentz-invariant state of zero energy and zero momentum. Considering that energy has been loosely thought to be essential to substance in modern physics, it seems that the vacuum could not be taken as a kind of substance. Here we run into a profound dilemma, which hints at the necessity of changing our conception of substance, and of energy being a substantial property.

In two respects, the mode of conveying interactions in QFT differs from that in the classical field program. First, interactions in QFT are realized by local couplings among field quanta, and the exact meaning of coupling here is the creation and annihilation of the quanta. Second, the actions are transmitted, not by a continuous field, but by discrete virtual particles that are locally coupled to real particles and propagate between them. Thus the description of interactions in QFT is deeply rooted in the concept of localized excitation of operator fields through the concept of local coupling. Yet the local excitation requires, owing to the uncertainty relation, that arbitrary amounts of momentum are available. The result of a localized excitation would not then be only a single momentum quantum, but rather a superposition of all appropriate combinations of momentum quanta. And this has significant consequences. First, the interaction is transmitted not by a single virtual momentum quantum, represented by an internal line in the Feynman diagram, but by a superposition of an infinite number of appropriate combinations of virtual quanta. This requirement is an entailment of the basic assumption of a field ontology in QFT. Second, the infinite number of virtual quanta with arbitrarily high momentum leads to infinite contributions from their interactions with real quanta. This is the famous divergence difficulty. Thus, QFT cannot be considered a consistent theory so long as this serious difficulty has not been resolved. Historically, the difficulty was first circumvented by a renormalization procedure.

The essence of the original renormalization procedure is the absorption of infinite physical quantities into the theoretical parameters of mass and charge. This is equivalent to blurring the exact-point model underlying the concept of localized excitations. Although quantum electrodynamics meets the requirement of renormalizability, Fermi's theory of weak interactions and the meson theory of the strong nuclear force both fail to do so.

Gauge invariance is a general principle for fixing the forms of fundamental interactions, on the basis of which a new program, the gauge field program, for fundamental interactions develops within the quantum field program. Gauge invariance requires the introduction of gauge potentials, whose quanta are responsible for transmitting interactions and for compensating the additional changes of internal degrees of freedom at different spacetime points. The role that gauge potentials play in gauge theory is parallel to the role that gravitational potentials play in GTR. Whereas the gravitational potentials in GTR are correlated with a geometrical structure (the linear connection in the tangent bundle), the gauge potentials are correlated with a similar type of geometrical structure, that is, the connection on the principal bundle.

Deep similarity in theoretical structures between GTR and gauge theory suggests the possibility that gauge theory may also be geometrical in nature. Recent developments in fundamental physics (supergravity, modern Kaluza-Klein theory, and string theory) have been proposed to associate gauge potentials with the geometrical structures in extra dimensions of spacetime (see chapter 16). Thus it seems reasonable to regard the gauge field program as a synthesis of the geometrical program with the quantum field program. If, that is, we express the gauge field program in such a way that interactions are realized through quantized gauge fields (whose quanta are coupled with material fields and are responsible for the transmission of interactions) that are inseparably correlated with a kind of geometrical structure existing either in internal space or in the extra dimensions of spacetime.

Let us return now to the role of the vacuum in quantum mechanics and QFT. The operator fields and field quantization, especially Jordan-Wigner quantization, had their direct physical interpretation only in terms of the vacuum state. Let us say a few words about the physicists' conception of the vacuum before and after the introduction of field quantization in the late 1920s. After Einstein had put an end to the concept of the ether, the field-free and matter-free vacuum was considered to be truly empty space. The situation had changed, however, with the introduction of quantum mechanics. From then onward, the vacuum became populated again. In quantum mechanics, the uncertainty relation for the number N of light quanta and the phase θ of the field amplitude, $\Delta N \Delta \theta \geq 1$, means that if N has a given value zero, then the field will show certain fluctuations about its average value, which is equal to zero.

The next step in populating the vacuum was that taken by Dirac. In his relativistic theory of electrons (1928), Dirac met with a serious difficulty, namely the existence of states of negative kinetic energy. As a possible solution to that difficulty, he proposed a new conception of the vacuum: all of the states of negative energy have already been occupied, and none of the states of positive energy are occupied (see chapter 1). But then, owing to Pauli's exclusion principle, the transition of an electron from the positive energy state to the negative energy state could not happen.

This vacuum state was not an empty state, but rather a sea filled with negative-energy electrons. The sea as a universal background was itself unobservable, yet a hole in the negative-energy sea was observable, behaving like a particle of positive energy and positive charge. Dirac took the hole as a new kind of particle with a positive charge and the mass of the electron. It is easy to see that Dirac's infinite sea of unobservable negative-energy electrons of 1930–31 was analogous to his infinite set of unobservable zero-energy photons of 1927. Such a vacuum, consisting of unobservable particles, allowed for a kind of creation and annihilation of particles. Given enough energy, a negative-energy electron can be lifted up into a positive-energy state, corresponding to the creation of a positron (the hole in the negative-energy sea) and an ordinary electron. And of course the reverse annihilation process can also occur.

One might say that all these processes can be accounted for by the concept of the transition between observable and unobservable states, without introducing the ideas of QFT. This is true. But at the same time, it is precisely the concepts of the vacuum and of the transition that have provided QFT with an ontological basis. In fact, the introduction of the negative-energy sea, as a solution to the difficulty posed by Dirac's relativistic theory of the electron, implied that no consistent relativistic theory of a single electron would be possible without involving an infinite-particle system, and that a QFT description was needed for the relativistic problems. Besides, the concept of transition between observable and unobservable states did indeed provide a prototype of the idea of excitation of QFT, and hence afforded the field operators a direct physical interpretation.

The concept of creation and destruction (annihilation) of particles and of their transition from one state (of negative energy) to another state (of positive energy) is maybe the most striking (and strangest) feature of Dirac's visionary theory of the vacuum. In fact, Dirac's idea of the vacuum as a kind of substratum shared some of its characteristic features with the ether model, which explains why he returned to the idea of an ether in his later years, although in his original treatment of creation and annihilation processes he did not explicitly appeal to an ether model.

One striking physical feature of Dirac's vacuum, similar to that of ether, is that it behaves like a polarizable medium. An external electromagnetic field distorts the simple electron-wave function of the negative-energy sea, and thereby produces a charge-current distribution acting to oppose the inducing field. As a consequence, the charges of particles would appear to be reduced. That is, the vacuum could be polarized by the electromagnetic field. The fluctuating densities of charge and current (which occur even in the electron-free vacuum state) can be seen as the electron-positron field counterpart of the fluctuations in the electromagnetic field.

In sum, the situation after the introduction of the filled vacuum is this. Suppose we begin with an electron-positron field Ψ. Such a field will create an accompanying electromagnetic field that reacts on the initial field Ψ and alters it. Similarly, an electromagnetic field will excite the electron-positron field Ψ, and the associated electric current acts upon and alters the initial electromagnetic field. The electromagnetic field and the electron-positron field are thus intimately connected, neither of them having a physical meaning independent of the other. What we have, then, is a coupled system consisting of electromagnetic field and electron-positron field, and the description of one physical particle is not to be written down *a priori*, but emerges only after a complicated dynamical problem has been solved.

The idea was that the vacuum was actually the scene of wild activities (fluctuations), in which infinite negative energy electrons existed, and it was mitigated later on by eliminating the notion of the actual presence of these electrons. Dirac's idea of the vacuum entailed a revolutionary meaning, and its notion of regarding the vacuum as not empty but substance-filled remains in our present conception of the vacuum. One may say that the conception of the substance-filled vacuum is strongly supported by the fact that the fluctuations of matter density in the vacuum remain even after the removal of the negative-energy electrons. This, as well as the electromagnetic vacuum fluctuations, is an additional property of the vacuum.

Here we run into the most profound ontological dilemma in QFT. On the one hand, according to special relativity, the vacuum must be a Lorentz invariant state of zero energy, zero momentum, zero angular momentum, zero charge, zero whatever. That is, a state of nothingness. Considering that energy and momentum have been thought to be the essential properties of substance in both modern physics and modern metaphysics, the vacuum cannot with confidence be regarded as a substance. But at the same time, the fluctuations found to exist in the vacuum strongly indicated that the vacuum must be something substantial, certainly not empty. A possible way out of the dilemma might be to redefine "substance" and to deprive energy and momentum of their accepted role as the defining properties of a substance. But this move would be too ad hoc, and could not find support from the lessons of other instances.

Another way out of the dilemma is to take the vacuum as a kind of pre-substance, an underlying substratum having a potential substantiality. It can be excited to become substance by energy and momentum, and become physical reality if various other properties are injected into it. The real particles come into existence only when we disturb the vacuum, by exciting it with energy and other properties. What shall we say about the local field operator whose direct physical meaning was thought to be the excitation of a physical particle? First of all, the localized excitation described by a local field operator $O(x)$ acting on the vacuum implies the injection of energy, momentum, and other special properties into the vacuum at a spacetime point. It also implies, owing to the uncertainty relations, that arbitrary amounts of energy and momentum are available for various physical processes. Evidently, the physical realization of these properties, symbolized by $O(x)|\text{vac}\rangle$, will not constitute a single particle state, but must be a superposition of all appropriate multi-particle states. For example, $\Psi_{el}(x)|\text{vac}\rangle = a|1 \text{ electron}\rangle + \Sigma a' |1 \text{ electron} + 1 \text{ photon}\rangle \Sigma a'' |1 \text{ electron} + 1 \text{ photon} + 1 \text{ electron}\rangle + \ldots$, where $\Psi_{el}(x)$ is the so-called dressed field operator, a^2 is the relative probability that a single bare (naked) particle state can be realized by the excitation $\Psi_{el}(x)$, and so on. As a result, the field operators no longer refer to the physical particles and become abstract dynamical variables, with the aid of which one constructs the physical state. Then how can we pass from the underlying dynamical variables (local field operators), with which the theory begins, to the observable particles? This is a task that is fulfilled temporally only with the aid of the renormalization procedure.

Let us roughly describe this procedure. In quantum field theory (QFT) and the statistical mechanics of fields, renormalization refers to a collection of techniques used to construct mathematical relationships or to approximate relationships between observable quantities, when the standard assumption that the parameters of the theory are finite breaks down, yielding the result that many observable quantities are infinite. Renormalization arose in quantum electrodynamics as a means of making sense of the infinite results of various calculations, and of extracting finite answers to properly posed physical questions.

When developing QED in the 1940s, S. Tomonaga, J. Schwinger, R. Feynman, and F. Dyson discovered that, in perturbative calculations, problems with divergent integrals abounded, and one way of describing the divergence in QED is that they appear as the consequences of calculations involving Feynman diagrams that entail closed loops of virtual particles. These diagrams appear in the perturbative approximation of QFT. Each looped diagram represents a perturbation of, or small correction to, a diagram without loops. Intuitively, diagrams with more and more loops should give smaller and smaller corrections to the values of diagrams that contain no loops. However, when the contributions of these loop diagrams are naively calculated, they become infinitely large. One type of loop would be exemplified by a situation in which a virtual electron-positron pair appears out of the vacuum, interacts with various photons, and then annihilates itself. Another would be an electron-photon interaction, as in fig. 7.

Although virtual particles obey conservation of energy and momentum, they can possess combinations of energies and momenta not allowed by the classical laws of motion; generally, physicists are not comfortable with that notion. Fur-

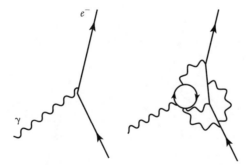

Fig. 7. Renormalization in quantum electrodynamics: The simple electron-photon interaction that determines the electron's charge at one renormalization point is revealed to consist of more complicated interactions at another such point.

thermore, whenever a loop appears, the particles involved in the loop are not individually constrained by the energies and momenta of incoming and outgoing particles, since a variation in, say, the energy of one particle in the loop can be balanced
by an equal and opposite variation in the energy of another particle in the loop.
Therefore, in order to calculate the contribution of a probability amplitude, one
must integrate over *all* possible combinations of energy and momentum in the
loop—and these integrals are often divergent, that is, they give infinite answers
(infinite mathematical terms). The most theoretically troublesome divergences
are the "ultraviolet" (UV) ones associated with large energies and momenta of the
virtual particles in the loop, or, equivalently, very short wavelengths and high
frequencies of the fields for which these particles are the quanta. These divergences are, therefore, fundamentally short-distance, short-time phenomena.

The diagram in fig. 8 shows one of the several one-loop contributions to
electron-electron scattering in QED. The electron on the left side of the diagram,
represented by the solid line, starts out with four-momentum p^μ and ends up
with four-momentum r^μ. It emits a virtual photon carrying $p^\mu - r^\mu$ to transfer
energy and momentum to the other electron. But before that happens, in this
diagram, it emits another virtual photon carrying four-momentum q^μ, and it
reabsorbs this one after emitting the other virtual photon. Energy and momentum conservation do not determine the four-momentum q^μ uniquely, so all possibilities contribute equally and we must integrate. This integral is divergent,
and in fact infinite unless we cut off at finite energy and momentum in some
way. Similar loop divergences occur in other quantum field theories.

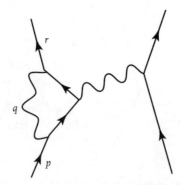

Fig. 8. A diagram contributing to electron-electron scattering in quantum electrodynamics. The loop has an ultraviolet divergence.

The solution was to realize that the quantities initially appearing in the theory's formulae (such as the formula for the Lagrangian), representing such things as the electron's electric charge and mass, as well as normalizations of the quantum fields themselves, did *not* actually correspond to the physical constants measured in the laboratory. As we have already said, they were *bare* quantities that did not take into account the contribution of virtual-particle loop effects *to the physical constants themselves*. Among other things, these effects would include the quantum counterpart of the electromagnetic back-reaction that so vexed classical theorists of electromagnetism. In general, these effects would be just as divergent as the amplitudes under study in the first place, and finite measured quantities would in general imply divergent bare quantities.

In order for us to make contact with reality, the formulae would have to be rewritten in terms of measurable, *renormalized* quantities. The charge of the electron, say, would be defined in terms of a quantity measured at a specific kinematic *renormalization point* or *subtraction point* (which will generally have a characteristic energy, called the renormalization scale, or simply the energy scale). The parts of the Lagrangian left over, involving the remaining portions of the bare quantities, could then be reinterpreted as *counterterms*, manifested in divergent diagrams exactly *canceling out* the troublesome divergences for other diagrams.

For example, in the Lagrangian of QED,

(6.1)
$$\mathcal{L} = \overline{\psi}_B [i\gamma_\mu (\partial^\mu + ie_B A_B^\mu) - m_B]\psi_B - 1/4 \, F_{B\mu\nu} F_B^{\mu\nu},$$

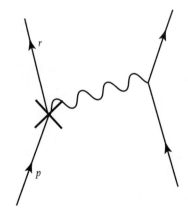

Fig. 9. The vertex corresponding to the Z_1 counterterm cancels the divergence in fig. 8.

the fields and coupling constant are really bare quantities, hence the subscript B above. Conventionally, the bare quantities are written so that the corresponding Lagrangian terms are multiples of the renormalized one:

$$(\overline{\psi}m\psi)_B = Z_0\, \overline{\psi}m\psi$$

(6.2) $$(\overline{\psi}(\partial^\mu + ieA^\mu)\psi)_B = Z_1\, \overline{\psi}(\partial^\mu + ieA^\mu)\psi$$

$$(F_{\mu\nu}F^{\mu\nu})_B = Z_3 F_{\mu\nu}F^{\mu\nu}.$$

A term in this Lagrangian, for example, the electron-photon interaction illustrated in fig. 9, can then be written as follows:

(6.3) $$\mathscr{L}_I = -e\, \overline{\psi}\gamma_\mu A^\mu \psi - (Z_1 - 1)e\, \overline{\psi}\gamma_\mu A^\mu \psi.$$

The physical constant e, the electron's charge, can then be defined in terms of some specific experiment; we set the renormalization scale equal to the energy characteristic of this experiment, and the first term gives the interaction we see in the laboratory (up to small, finite corrections from loop diagrams, providing such exotica as the high-order corrections to the magnetic moment). The rest is the counterterm. If we are lucky, the *divergent* parts of loop diagrams can all be decomposed into pieces with three or fewer legs, with an algebraic form that can be canceled out by the second term (or by the similar counterterms that come from Z_0 and Z_3). As a result, we have that QED theory is indeed *renormalizable*. The diagram in fig. 8, showing the Z_1 counterterms interaction vertex, cancels out the divergence from the loop in fig. 6.

Chapter 7

The Role of the Vacuum in Quantum Electrodynamics, the Casimir Effect, and Vacuum Polarization

Relativistic quantum field theory, as exemplified by quantum electrodynamics (QED), makes a very clear distinction between what we would intuitively understand to be an *absolute* void and what we experience as the vacuum of space. After all, even in the deepest reaches of space we will find an atom or molecule here and there, and photons of energy are flying through continuously at the speed of light, but there is also a *quantum potential* at *every point* in the vacuum of our three-dimensional physical space. Under the proper conditions, matter and energy can literally be made to materialize out of what we used to think of as nothing.

According to quantum mechanics, then, the vacuum is not empty, but teeming with virtual particles that constantly wink in and out of existence. One strange consequence of this sea of activity is the Casimir effect (1948). Two smooth flat metal surfaces inevitably attract one another if they are close enough. The Casimir force is so weak that it has rarely been detected at all. Recent work, however, has produced the most precise measurements of the phenomenon ever made, with the help of an atomic force microscope, which has the capacity to test the strangest aspects of the Casimir effect. The simplest explanation of the effect is that the two plates attract because their reflective surfaces exclude virtual photons of wavelengths greater than the separation distance. This reduces the energy density between the plates compared with the ambience, and—much the way external air pressure tends to collapse a slightly evacuated vessel—the Casimir force tends to pull the plates toward one another. But the most puzzling aspect of the theory is that the force depends on geometry—to be more precise, on the form of the plates. In fact, if they are replaced by hemispherical shells (bullets), the force is repulsive. Spherical surfaces somehow "enhance" the

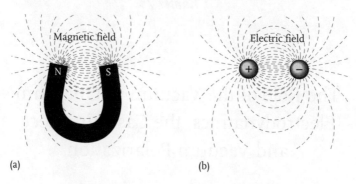

Fig. 10. Maxwell's equations give rise to magnetic field lines (a) of the same shape as electric field lines (b).

density of virtual photons. There is no simple or intuitive way to anticipate which way the force will go before carrying out the complicated calculations the situation requires.

The *electromagnetic* force has the particularity that it can propagate over long distances, and in consequence it is also experienced in the everyday world. For example, the British physicist James Clerk Maxwell had given the mathematical formulation of electrodynamics as early as 1873 (fig. 10).

The ideal particle for studying the effects of electromagnetic forces is the electron. It is a lepton, which means that it is insensitive to the strong force. Its mass is much smaller than that of most other particles; only later would people realize that this is precisely the reason why all sorts of indirect effects on the electron due to other particles can be neglected to a good approximation. The weak force does affect the electron, but its effects are so weak that they can safely be disregarded.

It is understandable why the fundamental theory for the interactions between electrons and photons was the first to be completed. The theory was called "quantum electrodynamics" (QED). The accuracy with which all sorts of properties of the electron could be calculated using the theory was impressive. One of the most striking examples of this was the calculation of the electron's magnetic dipole moment, μ. Because the electron rotates about its axis and is also electrically charged, it acts as a miniature (small) magnet. Paul A. M. Dirac, who was the first to write down a quantum mechanical equation for the electron that also agreed with relativity theory, found that the strength of this little magnet could be calculated in terms of known constants of nature, these being

Planck's constant, the speed of light, and the electric charge and mass of the electron:

(7.1) $\mu = e\hbar/4\pi m_e$.

But this equation still neglected certain indirect effects on electrons due to the photons surrounding them; these effects were later calculated by Julian Schwinger (1951) and others, with ever increasing accuracy.

Before this brilliant result could be obtained, however, some mountains had to be moved. An essential aspect of elementary particle theory is that particles can be both "created" and "annihilated." A consequence of this aspect is that the total number of particles involved in any interaction with an electron is constantly changing. There are two ways of contemplating this theory: on the one hand, we have become accustomed to describing all processes in terms of particles being created and annihilated in various places and times; but we can also view these processes as an ocean of waves crossing and affecting each other (an ocean that came to be called the "Dirac sea"). The waves of this ocean are said to be in "fields." Every particle type has its corresponding field. The photon belongs to the electromagnetic field. The electron has a very curious type of "electron field" (the "Dirac field"). In their wave-like oscillations these fields react upon each other's presence, and all of this is neatly governed by their "field equations." Quantum electrodynamics was the first system for which these equations came to be known, and so became the first prototype of a "quantum field theory." By 1930, physicists were already aware of the problem that had to be solved. Dirac realized the necessity to introduce the electron's antiparticle, the *positron*, which was discovered experimentally by Carl D. Anderson in 1932.

Quantum electrodynamics, for its part, is said to be a *renormalizable* theory. Roughly, this means the following. We have to start with the so-called "naked electron," that is, an electron with no photons in its vicinity. Actually, an electron cannot be separated from its surrounding photons, but we ignore this fact for the time being. This naked electron is given some value for its "naked" electric charge and its "naked" mass. If we try to calculate the magnetic moment for this naked electron we find, regrettably, that it is infinite. This is nonsense. But if we calculate what happens if photons enter the vicinity of this electron, we find that the photons create new electrons and positrons. This would not be noticed directly, but these additional particles have all sorts of effects on the electron. First, they act as a neutralizing shield (or guard) against the electric charge. This effect, called *vacuum polarization*, causes the electron's charge to change. If we try to calculate this change, alas (unfortunately), we find it too to be infinite. Second, the extra photons, electrons, and positrons have an effect on the electron's mass (they carry energy, hence also mass). This mass change also turns out to be infinite. In short: the true charge and the true mass of a real, "physical," electron, are quite different from those of the naked electron.

The replacement of the "naked" charge and mass of the electron by the physically observed values is called "renormalization." If we obey the rules following from this replacement carefully enough, we find, somewhat surprisingly, that all of those annoying infinities cancel out. An exact expression for the strength of the electron magnetic dipole moment remains to be determined. 1.001 159 652 19 ± 0.000 000 000 01 times the combination of constants first given by Dirac. Quantum electrodynamics gives for this number 1.001 159 652 17 ± 0.000 000 000 03. Ergo: the infinities are not in the electron or in the forces acting on the electron, but solely in our hypothetical naked electron. The naked mass and the naked charge are infinite (or, rather, ill-defined), but we can never observe these anyhow.

Chapter 8

Hole Theory,
Negative Energy Solutions,
and Vacuum Fluctuations

Notwithstanding the successes of the Dirac equation, we must face the interpretation of negative energy solutions. Their presence is difficult to accept, since in the final analysis they make all positive energy states unstable. A solution was proposed by Dirac as early as 1930a in terms of a many-particle theory. Although this will not stand as a final position, since for example it does not apply to scalar particles, it is instructive for us to retrace his thinking. His reasoning provides an intuitive physical picture useful in practical instances, and permits fruitful analogies with different situations, such as the electrons in a metal. Its major assumption is that all of the negative levels are filled up in the vacuum state. According to the Pauli exclusion principle, this prevents any electron from falling into the negative energy states, and thereby ensures the stability of positive-energy physical states. In turn, an electron of the negative energy sea may be excited to a positive energy state. It then leaves a hole in the sea. This hole in the negative energy, negatively charged states appears as a positive-energy, positively charged particle—the positron. Besides the properties of the positron, its charge $|e| = -e$ and its rest mass m_e, this theory also predicts two new observable phenomena:

First, the annihilation of an electron-positron pair. An electron (of positive energy) falls into a hole in the negative energy sea with the emission of radiation. From energy momentum conservation, at least two photons are emitted, unless a nucleus is present to absorb energy and momentum.

Conversely, an electron-positron pair may be created from the vacuum by an incident photon beam in the presence of a target, to balance energy and momentum. This is the process mentioned above; a hole is created while the excited electron acquires a positive energy.

Thus the theory predicts the existence of positrons that were in fact observed in 1932. Since positrons and electrons may be annihilated, we must abandon the interpretation of the Dirac equation as a wave equation. Further, the reason for discarding the Klein-Gordon equation no longer holds. The equation actually describes spinless particles, such as pions. The hole interpretation, however, is not satisfactory for bosons, since Fermi statistics play a crucial role in Dirac's argument. Even for fermions, the concept of an infinitely charged unobservable sea looks rather queer. We have instead to construct a true many-body theory to accommodate particles and antiparticles in a consistent way. This will be achieved by the "second quantization," i.e., the introduction of quantized fields capable of creating and annihilating particles.

Hole theory implies the existence of electrons and positrons, with the same mass and opposite charges, that obey the same equation. The Dirac equation must therefore admit a new symmetry corresponding to the interchange particle $\leftrightarrow$ antiparticle. We thus seek a transformation $\psi \to \psi^c$ reversing the charge, i.e., such that

(8.1)
$$(i\gamma^\mu \partial_\mu - e\gamma^\mu A_\mu - m)\psi = 0$$

and

(8.2)
$$(i\gamma^\mu \partial_\mu + e\gamma^\mu A_\mu - m)\psi^c = 0.$$

The vacuum fluctuations are beautifully evidenced by the Casimir effect. For Dirac fields, the stability of the vacuum and the exclusion principle lead to quantization according to anti-commutation rules. The original observation of Casimir (1948) is that, in the vacuum, the electromagnetic field does not really vanish but rather fluctuates. If we introduce macroscopic bodies—even uncharged—some work will be necessary to enforce appropriate boundary conditions. Because intuition on the sign of this effect is lacking, work here is meant in some algebraic sense, meaning the difference in zero-point energies between the two configurations. We may disregard the (infinite) contribution $\sum 1/2\hbar\omega_\infty$ to the Hamiltonian by arguing that it is observable. However, its variation can be measured.

Let us illustrate this point for the simple configuration of two large, parallel, perfectly conducting plates, as considered first by Casimir. Of course, we can study different geometries and different materials with similar results (except perhaps for crucial signs). We idealize the plates as two large parallel squares of size L at a mutual distance a (fig. 11) with $a \ll L$. Consider the energy per unit

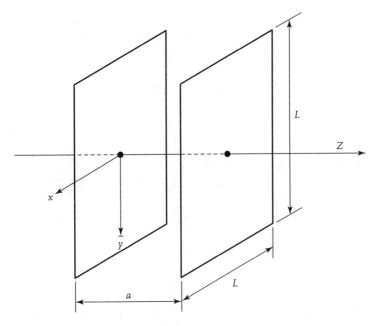

Fig. 11. The Casimir effect between two parallel plates.

surface of the conductor with respect to the vacuum. Its derivatives will be a force per unit surface with dimension $ML^{-1}T^{-2}$ (where M is mass, L length, and T time). The only quantities entering the problem are $\hbar$, c, and the separation a (the boundary conditions $\mathbf{E}$ perpendicular and $\mathbf{B}$ parallel to the plate at the interface do not introduce any dimensional quantity). Of course, the effect is proportional to $\hbar$, as is the zero-point energy. The force per unit surface is therefore proportional to $\hbar c/a^4$, the only quantity with the required dimension. We shall see that it is attractive.

Consider the modes inside the volume L^2a, where $L \gg a$ and we ignore the edge contributions. As we know, only transverse modes contribute to the energy. If the component k_z perpendicular to the plates is different from zero, it can take only discrete values $k_z = n\pi/a$ $(n = 1, 2, \dots)$ to allow for the nodes on the plates, and there are two polarization states. If, however, k_z vanishes, only one mode survives, and the zero-point energy of the configuration is

$$E = \sum 1/2\hbar\omega_\alpha = \hbar c/2 \sum_\alpha |\mathbf{k}_\alpha|$$

(8.3)

$$= \hbar c/2 \int L^2 d^2k_\| / (2\pi^2) \left[|\mathbf{k}_\|| + 2 \sum_{n=1}^{\infty} (k_\|^2 + n^2\pi^2/a^2)^{1/2} \right].$$

As it stands, this expression is, of course, meaningless, being infinite, but we must subtract the free value, which contributes in this same volume a quantity

(8.4)
$$E_0 = \hbar c/2 \int (L^2 d^2 k_\parallel / (2\pi)^2) \int_{-\infty}^{+\infty} (a dk_z/2\pi) 2\sqrt{k_\parallel^2 + k_z^2}$$

$$= \hbar c/2 \int (L^2 d^2 k_\parallel / (2\pi)^2) \int_0^\infty dn 2\sqrt{k_\parallel^2 + n^2 \pi^2/a^2}.$$

Therefore, the energy per unit surface is

(8.5)
$$\mathcal{E} = (E - E_0)/L^2 = \hbar c/2\pi \int_0^\infty k\,dk \Big(k/2 + \sum_{n=1}^\infty \sqrt{k^2 + n^2 \pi^2/a^2}$$

$$- \int_0^\infty dn \sqrt{k^2 + n^2 \pi^2/a^2} \Big).$$

This quantity is apparently still not defined, owing to ultraviolet (large k) divergences. But for wavelengths shorter than the atomic size, it is unrealistic to use a perfect-conductor approximation. Let us therefore introduce in the above integral a smooth cut-off function $f(k)$ equal to unity for $k < k_m$ and vanishing for $k \gg k_m$, where k_m is of the order of the inverse atomic size. Set $u = a^2 k^2/\pi$; then

(8.6)
$$\mathcal{E} = \hbar c (\pi^2/4a^3) \int_0^\infty du \Big[\sqrt{u/2} f(\pi/a\sqrt{u}) = \sum_1^\infty \sqrt{u + n^2} f(\pi/a\sqrt{u + n^2})$$

$$- \int_0^\infty dn \sqrt{u + n^2} f(\pi/a\sqrt{u + n^2}) \Big]$$

$$\mathcal{E} = \hbar c \pi/4a^3 \Big[1/2 F(0) + F(1) + F(2) + \dots - \int_0^\infty dn F(n) \Big].$$

Here we have defined

(8.7)
$$F(n) = \int_0^\infty du \sqrt{u + n^2} f(\pi/a\sqrt{u + n^2}).$$

The interchange of sums and integrals was justified, owing to the absolute convergence in the presence of the cut-off function. As $n \to \infty$, $F(n) \to 0$. We can use the Euler-MacLaurin formula to compute the difference between the sum and integral occurring in the above bracket:

(8.8) $1/2 F(0) + F(1) + F(2) + \ldots - \int_0^\infty dn F(n) = -1/2! \, B_2 F'(0) - 1/4! \, B_4 F'''(0) + \ldots$

The Bernoulli numbers B_ν are defined through the series

(8.9) $y/e^y - 1 = \sum_{\nu=0}^\infty B_\nu y^\nu / \nu!$

and $B_2 = 1/6$, $B_4 = -1/30$, . . . We have

(8.10) $F(n) = \int_{n^2}^\infty du \sqrt{u} f(\pi \sqrt{u/a})$ and $F(n) = -2n^2 f(\pi n/a)$.

We assume that $f(0) = 1$, while all its derivatives vanish at the origin, so that $F(0) = 0$, $F'''(0) = -4$, and higher derivatives of F are equal to zero. All reference to the cut-off has therefore disappeared from the final result

(8.11) $\mathcal{E} = (\hbar c \pi^2 / a^3)(B_4/4!) = -(\pi^2/720)(\hbar c/a^3)$.

The force per unit area $\mathcal{F}$ reads

(8.12) $\mathcal{F} = -(\pi^2/240)(\hbar c/a^4) = -(0.013/(a_{\mu m})^4) \, dyn/cm^2$,

and its sign corresponds to attraction.

This very tiny force has been demonstrated experimentally by Marcus Sparnaay (1958), who was able to observe both its magnitude and its dependence on the interplate distance! The above derivation may be criticized on the basis that we have seemingly disregarded the effects outside the plates. In the present case, they turn out to cancel exactly. The lesson is that vacuum fluctuations manifest themselves under circumstances quite different from those encountered in particle creation or absorption. By considering various types of bodies influencing the vacuum configuration, we may give an interesting interpretation of the forces acting on them.

Chapter 9

Further Theoretical Remarks on the Vacuum Fluctuations

Poincaré Conformal Invariance and Spontaneous Symmetry-Breaking Symmetry

In modern physics, the classical vacuum of still nothingness has been replaced by a quantum vacuum with fluctuations of measurable consequence. In fact, when we think about the vacuum in classical physics (say, in classical mechanics), we think of empty space unoccupied by any matter, through which particles can move unhindered, and in which fields are free from any of the nonlinear interaction effects which make, for example, electrodynamics in media so much more difficult. In Quantum Field Theory (QFT), the vacuum turns out to be quite different from this inert stage on which things happen; in fact, the vacuum itself is a nonlinear medium, a foamy bubble bath of virtual particles popping into and out of existence at every moment, a very active participant in the strange dance of elementary particles that we call the universe.

A metaphor that may make this idea a little clearer could be to think of the vacuum as a sheet of paper on which you write with your pen. Looked at on a large scale, the paper is merely a perfectly flat surface on which the pen moves unhindered. On a smaller scale, the paper is actually a tangle (or knot) of individual fibers arrayed in all directions, and against which the pen keeps hitting constantly, thus finding the necessary friction to allow efficient (and legible) writing. In the case where the paper is the vacuum, the analogue of the paper fibers is the bubbles and foams of virtual particle pairs that are constantly being created and annihilated in the quantum vacuum, the analogue of the pen is a particle moving through the vacuum, and the analogue of friction is the modification of the particle's behavior (as compared with the classical theory) that happens as a result of the particle interacting with virtual particle pairs.

At first sight, this description of the vacuum may seem like wild speculation, but it has in fact very observable consequences. In QED, the famous *Lamb shift* is a consequence of the interactions of the electron (in a hydrogen atom) with virtual photons, as are the *anomalous magnetic moment* of the electron and the scattering of light by light in the vacuum. In fact, none of the amazingly accurate predictions of QED would be realized without taking into account the effects of the quantum vacuum.

At the quantum level, we get Feynman diagrams entailing loops that describe how particles traveling through the quantum vacuum interact with virtual particles; the problem with these loops is that the virtual particles exist at very short distances, and hence can have very large momenta, by virtue of Heisenberg's uncertainty relation. At very large momenta, the deviation of the lattice theory from the continuum becomes quite evident, and the loops on the lattice thus contribute terms that differ considerably from what the same loops would contribute in the continuum. And then we find that this difference reintroduces the a-dependence that we got rid of classically by tuning our theory!

Quantum fluctuation is the temporary appearance of energetic particles out of nothing, as allowed by the uncertainty principle. It is synonymous with vacuum fluctuation. As mentioned before (see chapter 2), the uncertainty principle states that for a pair of conjugate variables such as position/momentum and energy/time, it is impossible to have a precisely determined value for each member of the pair at the same time. For example, a particle pair can pop out of the vacuum during a very short time interval. The uncertainty principle is illustrated by the schematic representation shown by fig. 12. An extension is applicable to the "uncertainty in time" and "uncertainty in energy" (including the rest mass energy mc^2). When the mass is very large (such as in the case of a

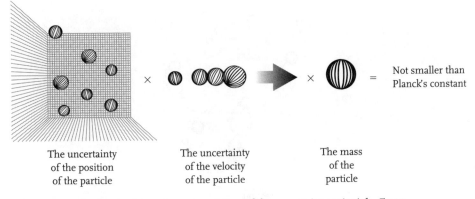

| The uncertainty of the position of the particle | The uncertainty of the velocity of the particle | The mass of the particle |

Fig. 12. A schematic representation of the uncertainty principle. From Heisenberg, 1927.

macroscopic object), the uncertainties and thus the quantum effect, become very small, and classical physics is applicable once more.

In classical physics (applicable to macroscopic phenomena), empty space-time is called the vacuum; and the classical vacuum is utterly featureless. In quantum mechanics, however, the vacuum is a much more complex entity. It is far from featureless and far from empty. The quantum vacuum is just one particular state of a quantum field (corresponding to some particles). It is the quantum-mechanical state in which no field quanta are excited, that is, no particles are present. Hence, it is the "ground state" of the quantum field, the state of minimum energy. Fig. 13 illustrates the kind of activities going on in a quantum vacuum. It shows particle pairs appearing, leading a brief existence, and then annihilating one another in accordance with the uncertainty principle.

Heisenberg's uncertainty principle applied to energy E and time t states that $\Delta E \, \Delta t \geq \hbar/2$. Then if Δt is sufficiently small, ΔE can become large enough to allow the appearance of virtual particle-antiparticle pairs. In the quantum vacuum, virtual particle-antiparticle pairs are continuously created and annihilated: the quantum vacuum is full of activity. The classical ("ordinary") vacuum, by contrast, is defined as the absence of matter: the classical vacuum is empty. (See figs. 14–17.)

First, the quantum vacuum can be thought of as a medium much like an ordinary gas. Its structure, however, is normally invisible, much like the water

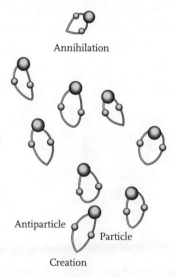

Annihilation

Antiparticle

Particle

Creation

Fig. 13. Apparition of particle pairs in a quantum vacuum.

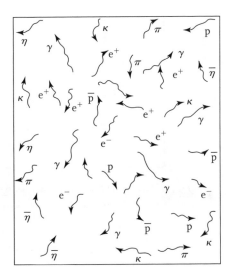

Fig. 14. "Seeing" the classical vacuum.   *Fig. 15.* "Seeing" the quantum vacuum.

of a transparent pond. To "see" the structure, one must perturb the medium: in the case of the transparent water, one could evidence the water by throwing a stone (a "disturbance") into the pond.

Second, when studying the quantum vacuum, the idea is to set up a disturbance with an external electromagnetic field, and to use a "probe" of some sort to investigate possible changes in the structure. A light beam could be such a probe.

Third, In the PVLAS experiment, the disturbance is introduced by a high-intensity (up to 6.6 *T*) magnetic field, and the probe is a linearly polarized laser beam: the properties of the quantum vacuum are recorded in the polarization state of the probe light, which has changed from linear to elliptical. This phenomenon is also called *vacuum magnetic birefringence*.

In quantum field theory, the vacuum state (also called the vacuum) is the quantum state having the lowest possible energy. By definition, it contains no physical particles. The term "zero-point field" is sometimes used as a synonym for the vacuum state of an individual quantized field. If the quantum field theory can be accurately described through perturbation theory, then the properties of the vacuum are analogous to the properties of the ground state of a quantum-mechanical harmonic oscillator (or more accurately, the ground state of a QM problem). In this case, the vacuum expectation value (VEV) of any field operator vanishes—the VEV of a field ϕ, which should be written as $\langle 0|\phi|0\rangle$, is usually

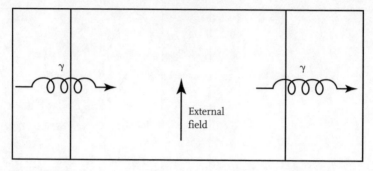

Fig. 16. In classical vacuum, there is no effect: the perturbing field and the probe light do not "mix," and the exiting probe photons are unchanged.

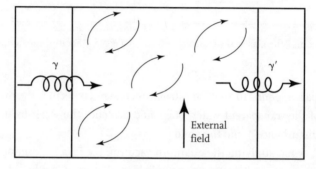

Fig. 17. In quantum vacuum, the perturbing field "changes" the structure of the quantum vacuum: the probe light and field now "mix," and the exiting photons carry information on the structure of the vacuum.

condensed to $\langle\phi\rangle$. For quantum field theories in which perturbation theory breaks down at low energies (for example, quantum chromodynamics [QCD] or the BCS theory of superconductivity), field operators may have nonvanishing vacuum expectation values called condensates. In the Standard Model of particle physics, the nonzero vacuum expectation value of the Higgs field, arising from spontaneous symmetry breaking, is the mechanism by which the other fields in the theory acquire mass.

In many situations, the vacuum state can be defined to have zero energy, although the actual situation is considerably more subtle. The vacuum state, written as $|0\rangle$ or $|\rangle$, is associated with a zero-point energy, and this zero-point energy has measurable effects. In the laboratory, it can be detected as the Casimir effect. In physical cosmology, the energy of the vacuum state appears as the cosmological constant. An outstanding requirement imposed on a potential Grand Unified

Theory is that the vacuum energy of the vacuum state must explain the physically observed cosmological constant.

For a relativistic field theory, the vacuum is Poincaré invariant. Poincaré invariance implies that only scalar combinations of field operators have nonvanishing VEVs. The VEV may break some of the internal symmetries of the Lagrangian of the field theory. In this case, the vacuum (that is, the ground state of the QFT) has less symmetry than the theory allows, and one says that spontaneous symmetry breaking has occurred. This means that, when the Hamiltonian (or the Lagrangian) of a system has a certain symmetry, but the ground state (i.e., the vacuum) does not, one says that spontaneous symmetry breaking (SSB) has taken place. When a continuous symmetry is spontaneously broken, massless gauge bosons (messenger particles that mediate the interactions) appear, corresponding to the remaining symmetry. This is called the Goldstone phenomenon, and the bosons are called Goldstone bosons. In other words, spontaneous symmetry breaking in physics takes place when a system that is symmetric with respect to some symmetry group goes into a vacuum state that is not symmetric. At this point, the system no longer appears to behave in a symmetric manner. It is a phenomenon that occurs naturally in many situations. The symmetry group can be discrete, such as the space group of a crystal, or continuous (i.e., a Lie group), such as the rotational symmetry of space.

A common example to help explain this phenomenon is a ball sitting on top of a hill. This ball is in a completely symmetric state, but it is not a stable one: the ball can easily roll down the hill. At some point, the ball will spontaneously roll down the hill, in one direction or another. The symmetry has been broken because the direction taken by the ball has now been singled out from other directions.

Consider now a well-known (and very nice) mathematical example: the "Mexican hat potential." In the simplest example, the spontaneously broken field is described by a scalar field. In physics, one way of seeing spontaneous symmetry breaking is through the use of Lagrangians. Lagrangians, which essentially dictate how a system will behave (evolve in time), can be split up into kinetic and potential terms

(9.1) $$\mathcal{L} = \partial^{\mu}\phi\,\partial_{\mu}\phi - V(\phi).$$

It is in this potential term ($V(\phi)$) that the action of symmetry breaking occurs. An example of a potential is illustrated in the graph below

(9.2) $$V(\phi) = -10|\phi|^2 + |\phi|^4.$$

Graph of spontaneous symmetry-breaking functions in equation (9.2).

This potential has many possible minima (vacuum states) given by

$$(9.3) \qquad\qquad \phi = \sqrt{5}\, e^{i\theta}$$

for any real θ between 0 and 2π. The system also has an unstable vacuum state corresponding to $\phi = 0$. This state has a $U(1)$ symmetry, but once the system falls into a specific stable vacuum state (corresponding to a choice of ϕ), this symmetry will be lost or spontaneously broken.

In the following, we offer some particularly interesting examples of spontaneously broken symmetry, which are taken from various fields of physics.

1. For ferromagnetic materials, the laws describing it are invariant under spatial rotations. Here, the order parameter is the magnetization, which measures the magnetic dipole density. Above the Curie temperature, the order parameter is zero, which is spatially invariant and there is no symmetry breaking. Below the Curie temperature, however, the magnetization acquires a constant nonzero value that points in a certain direction. (This holds in the idealized situation where we have full equilibrium; otherwise, translational symmetry gets broken as well.) The residual rotational symmetries that leave the orientation of this vector invariant remain unbroken, but the other rotations get spontaneously broken.

2. The laws describing a solid are invariant under the full Euclidean group, but the solid itself spontaneously breaks this group down to a space group. The displacement and the orientation are order parameters.

3. The laws of physics are spatially invariant, but here on the surface of the Earth, we have a background gravitational field (which plays the role of the order parameter here) which points downwards, breaking the full rotational symmetry.

This explains why up, down, and the horizontal directions are all "different" but all the horizontal directions remain isotropic.

4. General relativity has a Lorentz gauge symmetry, but in Friedmann-Robertson-Walker cosmological models, the mean 4-velocity field defined by averaging over the velocities of the galaxies (the galaxies act like gas particles at cosmological scales) acts as an order parameter breaking this Lorentz symmetry. Similar comment can be made about the cosmic microwave background.

5. Here on Earth, Galilean invariance (in nonrelativistic approximation) is broken by the velocity field of the Earth/atmosphere, which acts as the order parameter here. This explains why people before Galileo thought that moving bodies tend toward rest. We tend not to be aware of broken symmetries.

6. For the electroweak model, the Higgs field acts as the order parameter breaking the electroweak gauge symmetry to the electromagnetic gauge symmetry. Like the ferromagnetic example, there is a phase transition at the electroweak temperature.

7. For superconductors, there is a collective condensed matter field ψ which acts as the order parameter breaking the electromagnetic gauge symmetry.

8. In general relativity, diffeomorphism covariance is broken by the nonzero order parameter, the metric tensor field.

If a vacuum state obeys the initial symmetry, then the system is said to be in the Wigner mode; otherwise, it is in the Goldstone mode. In the Standard Model, spontaneous symmetry breaking (accomplished by using the Higgs boson) is responsible for the masses of the W and Z bosons. Recall that the Higgs boson particle is one quantum component of the theoretical Higgs field. In empty space, the Higgs field has an amplitude different from zero. This field, also known as a "nonzero vacuum expectation value," illustrates the concept that there is no such thing as a completely "empty" space. The existence of this nonzero vacuum expectation plays a fundamental role: it gives mass to every elementary particle that has mass, including the Higgs boson itself; in other words, it is a hypothetical massive scalar elementary particle that helps explain how otherwise massless elementary particles still manage to construct mass in matter. In particular, the acquisition of a nonzero vacuum expectation value spontaneously breaks electroweak gauge symmetry, often referred to as the Higgs mechanism. This is the simplest mechanism capable of giving mass to the gauge bosons while remaining compatible with gauge theories.

In essence, this field is analogous to a pool of molasses that "sticks" to the otherwise massless fundamental particles that travel through the field, converting them into particles with mass that forms the basis of the atom. In the Standard Model, the Higgs mechanism consists of two neutral and two charged component fields. Both of the charged components and one of the neutral fields are Goldstone bosons, which are massless and act as the longitudinal third-polarization components of the massive W^+, W^-, and Z bosons. The quantum of the remaining neutral component corresponds to the massive Higgs boson. Since the Higgs field is a scalar field, the Higgs boson has no spin and has no intrinsic angular momentum.

Vacuum energy is an underlying background energy that exists in space even when devoid of matter (then known as free space). The vacuum energy results in the existence of most (if not all) of the fundamental forces—and thus in all effects involving these forces, too. It is observed in various experiments, and it is thought (but not yet demonstrated) to have consequences for the behavior of the Universe on cosmological scales, where the vacuum energy is expected to contribute to the cosmological constant[1], which affects the expansion of the universe.

Vacuum energy has a number of consequences. For one, vacuum fluctuations are always created as particle/antiparticle pairs. The creation of these "virtual particles" near the event horizon of a black hole has been hypothesized by the physicist Stephen Hawking to be a mechanism for the eventual "evaporation" of black holes. The net energy of the universe remains zero so long as the particle pairs annihilate each other within Planck time. If one of the pair is pulled into the black hole before this, then the other particle becomes "real" and energy/mass is essentially radiated into space from the black hole. This loss is cumulative and could result in the black hole's disappearance over time; the time required is dependent on the mass of the black hole, but could be on the order of 10^{100} years for large solar-mass black holes. The grand unification theory (GUT) predicts a nonzero cosmological constant from the energy of vacuum fluctuations[2].

More Intuitive Remarks on the Casimir Effect and Force, and on Their Significance

Quantum mechanics was one of the most outstanding physical theories of the twentieth century. The other was the general theory of relativity. Quantum mechanics revealed nature at the microscopic scale as a strange and fascinating scientific object that seems to defy the natural human intuition developed through our everyday experience. Though this quantum character is present everywhere as the building block of the physical world, it seldom reveals itself at a macroscopic level.

The Casimir effect is one of those rare exceptions where a quantum effect can have significant consequences beyond the atomic level. More precisely, the Casimir effect is an outcome of quantum field theory, which states that all of the various fundamental fields, such as the electromagnetic field, must be quantized at each and every point in space.[1] In a naïve sense, a *field* in physics may be envisioned as if space were filled with interconnected vibrating balls and springs, and the *strength* of the field can be visualized as the displacement of a ball from its rest position. Vibrations in this field propagate, and are governed by the appropriate wave equation for the particular field in question. The second quantization of quantum field theory[2] requires that each such ball-spring combination be quantized, that is, that the strength of the field be quantized at each point in space. Canonically, the field at each point in space is a simple harmonic oscillator, and its quantization places a quantum harmonic oscillator at each point. Excitations of the field correspond to the elementary particles of the particle physics. But even the vacuum has a vastly complex structure. All calculations of quantum field theory must be made in relation to this model of the vacuum.

The vacuum has, implicitly, all the properties that a particle may have: spin, or polarization in the case of light, energy, and so on. On average, all of these properties cancel out: the vacuum is, after all, "empty" in this sense. One important exception is the vacuum energy or the vacuum expectation value of the energy. The quantization of a simple harmonic oscillator states that the lowest possible energy or zero-point energy that such an oscillator may have is $E = 1/2\hbar\omega$. Summing over all possible oscillators at all points in space yields an infinite quantity. To remove this infinity, one may assume that only differences in energy are physically measurable, much as the concept of potential energy has been treated in classical mechanics for centuries. This argument is in fact the underpinning of the theory of renormalization. In all practical calculations, this is how infinity is always handled. In a deeper sense, however, renormalization is unsatisfying, and the removal of this infinity presents a challenge in the search for a Grand Unification Theory (GUT). Currently, there is no compelling explanation for why this infinity should be treated as essentially zero; a nonzero value is essentially the cosmological constant, and any large value causes trouble in cosmology.

The simplest form of the Casimir effect, predicted in 1948 by H. B. G. Casimir, consists in the attraction between a pair of neutral, parallel conducting plates placed in the vacuum. This attractive force, purely quantum in origin, cannot be obtained using the classical description of an electromagnetic field,

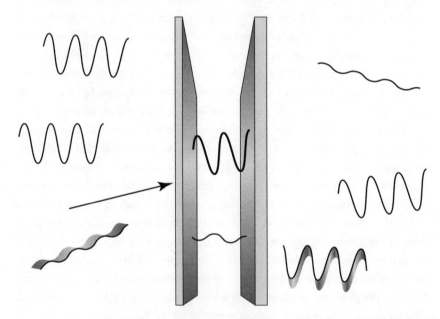

Fig. 18. *The Casimir effect.* The existence of ground-state fluctuation has been confirmed experimentally by the Casimir effect, a slight force between parallel metal plates. (*Left*) The reduced number of wavelengths that can fit between the plates. (*Right*) The wavelengths outside the confines of the plates.

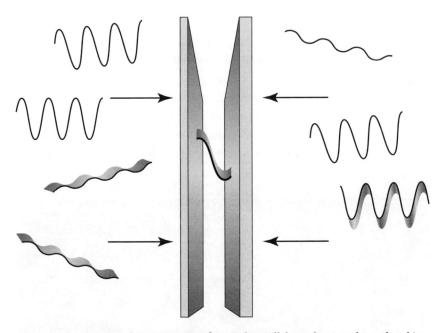

Fig. 19. The attraction between a pair of neutral, parallel, conducting plates placed in the vacuum. (*Left*) The energy density of ground-state fluctuations between the plates is less than the density outside, causing the plates to draw together. (*Right*) The energy density of ground-state fluctuations is greater outside the plates.

since it is a direct consequence of the existence of *zero-point fluctuations*: a turmoil of virtual particles that come in and out of existence and can violate the energy-momentum conservation in the system for very short periods of time, as described by Heisenberg's uncertainty principle. According to that principle, the more precisely the position of a subatomic particle is determined, the less precisely the momentum of the same particle is known in this instant, and vice versa.[3] The fluctuating virtual particles exert a "radiation pressure" on the plates that on average is greater outside the plates than between them, as shown in figs. 18 through 20. Though the Casimir effect is expected to hold for any type of quantum field, in most cases this phenomenon is considered to hold for the electromagnetic field, in view of the fact that this is the strongest fundamental interaction and is the most likely to generate measurable effects. It should be noticed that the Casimir phenomenon is now often used in a wide range of effects related to zero-point fluctuations, which extend far beyond the Casimir effect. In fact, zero-point fluctuations exist everywhere and are responsible for an infinite amount of energy in the universe. Hence, they may be expected to have vast bearing (influence) in many scientific problems.

An important characteristic of the Casimir effect is that it is a macroscopic quantum effect. For 2-plane parallel metallic plates of area $1\,cm^2$ separated by a great distance (on the atomic scale) of 1 micron, the value of the attractive force is approximately $10^{-7}\,N$. This force, though quite small, is now within the range of modern laboratory force. In fact, because the strength of the Casimir force falls off rapidly with distance, it is measurable only when the distance between the objects is extremely small. On a submicron scale, this force becomes so strong that it becomes the dominant force between uncharged conductors. Indeed, at

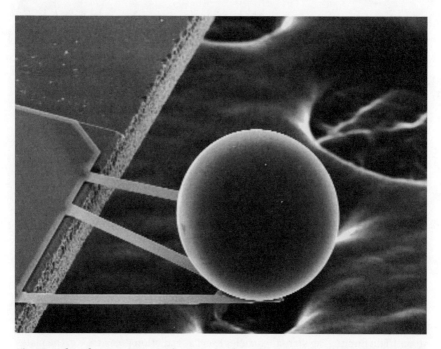

Fig. 20. A force from empty space: the Casimir effect. This tiny ball offers evidence that the universe will expand forever. Measuring slightly over one-tenth of a millimeter, the ball moves toward a smooth plate in response to energy fluctuations in the vacuum of empty space. The attraction is known as the Casimir effect. Today, evidence is accumulating that most of the energy density in the universe is in an unknown form dubbed dark energy. The form and genesis of dark energy are almost completely unknown but are postulated as relating to vacuum fluctuations similar to the Casimir effect, but generated somehow by space itself. This vast and mysterious dark energy appears to repel gravitationally all matter, and hence will likely cause the universe to expand forever. Understanding vacuum fluctuations is on the forefront of research seeking not only to better understand our universe but also to stop micro-mechanical machine parts from sticking together. By permission of Umar Mohideen.

separations of 10 nm—about a hundred times the typical size of an atom—the Casimir effect produces the equivalent of 1 atmosphere of pressure (101.3 *kPa*).

Although the Casimir force seems completely counterintuitive, it is actually well understood. In the old days of classical mechanics, the idea of a vacuum was simple. The vacuum was what remained if you emptied a container of all its particles and brought the temperature down to absolute zero. The arrival of quantum mechanics, however, completely changed our notion of a vacuum. All fields—in particular, electromagnetic fields—have fluctuations. In other words, at any given moment their actual value varies around a constant, mean value. Even a perfect vacuum at absolute zero has fluctuating fields known as "vacuum fluctuations," the mean energy of which corresponds to half the energy of a photon.

Vacuum fluctuations, however, are not some abstraction of a physicist's mind. They have observable consequences that can be directly visualized in experiments on a microscopic scale. For example, an atom in an excited state will not remain there infinitely long, but will return to its ground state by spontaneously emitting a photon. This phenomenon is a consequence of vacuum fluctuations. Imagine trying to hold a pencil upright on the end of your finger. It will stay there if your hand is perfectly stable and nothing perturbs the equilibrium. But the slightest perturbation will make the pencil fall into a more stable equilibrium position. Similarly, vacuum fluctuations cause an excited atom to fall into its ground state.

The Casimir force is the most famous mechanical effect of vacuum fluctuations. Consider the gap between two plane mirrors as a cavity. All electromagnetic fields have a characteristic "spectrum" consisting of many different frequencies. In a free vacuum, all of the frequencies are of equal importance. But inside a cavity, where the field is reflected back and forth between the mirrors, the situation is different. The field is amplified if integer multiples of half a wavelength can fit exactly inside the cavity. This wavelength corresponds to a "cavity resonance." At other wavelengths, in contrast, the field is suppressed. Vacuum fluctuations are suppressed or enhanced depending on whether their frequency corresponds to a cavity resonance or does not.

An important physical quantity when discussing the Casimir force is the "field radiation pressure." Every field—even the vacuum field—carries energy. Because all electromagnetic fields can propagate in space, they also exert pressure on surfaces, just as a flowing river pushes on a floodgate. This radiation pressure increases with the energy—and hence the frequency—of the electromagnetic field. At a cavity-resonance frequency, the radiation pressure is stronger inside the cavity than outside, and the mirrors are therefore pushed apart. Out of resonance, in contrast, the radiation pressure inside the cavity is smaller than that outside, and the mirrors are drawn toward each other.

Chapter 11

Dynamical Principles of Invariance and the Physical Interactions

Intuitively, a symmetry principle is a statement that the laws of nature continue to look the same when we change our point of view in some way. The special principle of relativity says that the laws of nature appear the same to observers moving with any constant velocity. There are other space-time symmetries, some of them telling us that the laws of nature look the same when we rotate or translate our laboratory, or reset our clocks. The "standard model," which corresponds to the group $SU(3) \times SU(2) \times U(1)$, is based on a set of symmetry principles, including the symmetries of space-time, as well as other symmetries called "internal symmetries." These require that the laws of nature take the same form when we make certain changes in the ways that the fields of the theory are labeled.

The concept of *renormalizability* arose from the effort in the late 1940s to make sense of the infinite energies and reaction rates that appeared when calculations in quantum electrodynamics were pushed beyond the first approximation. A theory is said to be renormalizable if these infinities can all be canceled by suitable redefinitions of a finite number of parameters of the theory, such as masses and electric charges; that is the case for electrodynamics and for weak and strong interactions in the standard model of elementary-particle physics. But for gravity, renormalization theory fails, because of the nature of the inherent nonlinearities in general relativity.

In his 1909 book *The Theory of the Electron*, Hendrik Antoon Lorentz took the opportunity to comment on the difference between Einstein's special theory of relativity, proposed four years earlier, and his own work. Lorentz had tried to use an electromagnetic theory of electron structure to show that matter composed of electrons would when in motion behave in such a way as to make it impossible for us to detect the effects of its motion on the speed of light, thus explaining the persistent failure to detect any difference in the speed of light along, or at right angles to, the Earth's motion around the Sun. Einstein, in contrast, had taken it as a fundamental axiom that the speed of light is the same to all observers. From a modern point of view, what Einstein had done was to introduce a principle of symmetry—the invariance of laws of nature under changes in the velocity of the observer—as one of the fundamental laws of nature (Weinberg, 1978).

Since Einstein's time, we have become more and more familiar with the idea that symmetry principles of various sorts are legitimate fundamental hypotheses. The "standard model" (which achieved the unification of weak and electromagnetic interactions) is largely based on a set of assumed symmetry principles, and string theory can also be viewed in this way. But in the days of Lorentz and Einstein, symmetries were generally regarded as mathematical curiosities, of great value to crystallographers, but hardly worthy to be included among the fundamental laws of nature. It is not surprising that Lorentz was uncomfortable with Einstein's hypothesis of the principle of relativity.

This point, important both scientifically and philosophically, led to most fundamental developments in the past and current attempts to understand the structures of space-time, as well as to the way in which these structures are embodied in physical phenomena. Therefore, it deserves much discussion here. Symmetry and invariance considerations, and even conservation laws, undoubtedly played an important role in the thinking of the early physicists, such as Galileo and Newton, and probably even before them. But these considerations were not deemed to be particularly important and were articulated only rarely. Newton's equations were not formulated in any special coordinate system and thus left all directions and all points in space equivalent. They were invariant under rotations and displacements, as we say now (see Arnold [1978]). There was little point in emphasizing this fact, or in conjuring up the possibility of laws of nature that show a lower symmetry.

In recent times, this situation has changed radically, as far as the invariance of the equations is concerned, principally as a result of Einstein's theories. Einstein considered the postulates about the symmetry of space, that is, the equivalence of directions and of different points of space, as basic to any inquiry of the physical laws. He also reestablished, in a modified form, the equivalence of coordinate systems in motion and at rest. Einstein's great advance in 1905 was to put symmetry first, to regard the symmetry principle as the primary feature of nature constraining the allowable dynamical laws. The universal principle proposed by Einstein in 1905 prescribes that the laws of physics are invariant under Lorentz transformation, when going from one inertial system to another (arbitrarily chosen) inertial system. This follows from the joint assertion of the thoroughgoing physical equivalence of all inertial frames, assumed by the Relativity Principle, and the constancy of the speed of light in vacuo, regardless of the state of motion of its source. Moreover, the invariance of the relationships that constitute the laws of electrodynamics (i.e., Maxwell's equations for the electromagnetic field) were seen to be invariant under a Lorentz transformation, in the sense that these equations take the same form when expressed in any set of frames related by this transformation. Stated otherwise, Maxwell's equations may be derived from relativistic invariance. In Einstein's general relativity theory, the principle of equivalence—a principle of local symmetry (the invariance of the laws of nature under local changes in the space-time coordinates)— dictated the dynamics of gravity, of space-time itself. As far as the conservation laws are concerned, their significance became evident when, as a result of the interest in Bohr's atomic model in 1913, the angular-momentum conservation theorem became all-important. But what was already established in the early 1900s was the connection between the conservation laws and the fundamental symmetries of space and time.

Since the turn of the century, the link between conservation laws and invariance principles has been generally accepted. In addition, the concept of symmetry and invariance has been extended into a new area, an area where its roots are much less close to direct experience and observation than they had been in the classical area of space-time symmetry. It may be useful, therefore, to discuss first the relations of phenomena, laws of nature, and invariance principles to each other. This relation is not quite the same for the classical invariance principles, which will be called *geometrical*, as it is to the new principle, which will be called *dynamical*. Another point to be recognized here is the relation between conservation laws and invariance principles. What is worth noting is the similarity between the relation of the laws of nature to physical events, on the one hand, and the relation of the invariance principles to the laws of nature, on the other hand. The similarity implies that the invariance principles provide structure or coherence to the laws of nature, just as the laws of nature provide structure and coherence to the set of physical events (see Wigner [1967] and Yang [1983]).

What, then, is the difference between the old and well-established geometrical principles of invariance and the more novel, dynamical principle? The geometrical principles of invariance, though they give a structure to the laws of nature, are formulated in terms of events themselves. Thus, the time-displacement invariance (in special relativity), when properly formulated, is this:

the correlations between events depend only on the time intervals between the events, not on the time at which they first take place. If P_1, P_2, P_3 are positions that a planet, say, can assume at times t_1, t_2, t_3, the planet could also assume these positions at times $t_1 + t$, $t_2 + t$, $t_3 + t$, where t can be quite arbitrary. At the same time, the new, dynamical principles of invariance are formulated in terms of the laws of nature. They apply to specific types of interactions, rather than to any correlation between events. Thus, we say that the electromagnetic interaction is gauge-invariant, with reference to a specific law of nature that regulates the generation of the electromagnetic field by charges, and the influence of the electromagnetic field on the motion of the charges. (On this subject, see Salam [1990].)

It follows that the dynamical types of invariance are based on the existence of specific kinds of interactions. We now recognize three distinct kinds of fundamental interactions[1]: the gravitational force (which is by far the weakest of the known interactions), the unified electroweak interaction[2] (which is responsible for beta decay, the decay of the *m* meson, and similar phenomena), and the strong interaction[3] (which varies differently with distance, and holds only inside the atomic nucleus). Thus, we have given up, at least temporarily, the hope of a single basic interaction. Furthermore, every interaction has a dynamical invariance group, such as that for the electromagnetic interaction. Some physicists have stimulated a fruitful line of thinking about how the interaction itself might be understood once its group is known. But the groups seem to be quite disjointed, and there seems to be no connection among the various groups that characterize the interactions, or between these groups and the geometrical symmetry group, which is a single, well-defined group with which we have been familiar for many years.

Because the geometrical principles of invariance were first recognized by Poincaré, it seems natural to call the group formed by these invariant properties the *Poincaré group*. Only Einstein, in his special theory of relativity (as mentioned previously), brought out the true meaning and importance of these principles. The group contains, first, displacements in space and time. This means that the correlations between events are the same everywhere and at all times, that the laws of nature are the same no matter when and where they are established.

In a series of papers on atomic structure and molecular spectra, written between 1926 and 1928, E. Wigner laid the foundation both for the application of group theory to quantum mechanics and for the role of symmetry principles in quantum mechanics. Toward the end of the 1930s, he turned his attention to time-dependent symmetries, the invariance groups that include time-translation invariance. In Wigner (1939), he considered the Poincaré group, which had not been seriously studied previously by either mathematicians or physicists, and posed two questions: what are the unitary representations of the Poincaré group, and what is their physical significance? The answers to these questions consist in showing that in relativistic quantum mechanics, Poincaré transformations were represented by faithful representations of their covering group, $ISL(2, \mathbf{C})$, which is not compact, and its unitary representations are infinite representations. This allowed him to build a complete classification and an explicit construction of all the irreducible representations. He then offered a *definition* of what we mean by an elementary particle, which according to Wigner should be identified as an irreducible representation of the Poincaré group.

It is helpful at this point to emphasize the fact that the laws of nature, that is, the correlations between events, are the entities to which the symmetry laws, not the events themselves, apply. Naturally, the events vary from place to place, but if one observes the positions of a thrown rock at three different times, one will find a relation between those times and positions, and this relation will be the same at all points on the Earth.

The second symmetry, not at all as obvious as the first one, postulates the equivalence of all directions. In other words, the events between which the laws of nature establish correlations are not simply the three positions of the thrown rock, but the three positions of the rock with respect to the Earth.

The last symmetry—the independence of the laws of nature from the state of motion in which these systems are observed, so long as motion is uniform—is not at all obvious. One of its consequences is that the laws of nature determine not the velocity but the acceleration of a body: whereas the velocity differs in coordinate systems moving with different speeds, the acceleration is the same as long as the motions of the coordinate systems are uniform with respect to each other.

The conservation laws for energy, and for linear and angular momentum, are direct consequences of the symmetries just enumerated. This is most evident in quantum-mechanical theory, where they follow directly from the kinematics of the theory, without making use of any dynamical law, such as the Schrödinger equation. From this point of view, classical theory is a limiting case of quantum theory. Hence, any equation valid in quantum theory, for any value of Planck's constant $\hbar$, is valid also in the limit $\hbar = 0$. There is thus a clear connection between conservation laws and space-time symmetry in classical physics. In quantum mechanics, the conservation laws follow already from the basic kinematics concepts. The point is simply that the states in quantum mechanics are vectors in an abstract space, and the physical quantities, such as position, momentum, etc., are operators on these vectors. It then follows, for instance, from the rotational invariance that, given any state ϕ, there is another state ϕ_α that looks just like ϕ in the coordinate system obtained by a rotation α about the Z-axis. Actually, the angular momentum about the Z-axis is the limit of $(1/\alpha)(Z_\alpha - 1)$ for infinitely small α.

The other conservation laws are derived in the same fashion. Furthermore, the transformation operators, or at least the infinitesimal ones among them, play a double role and are themselves the conserved quantities.

With the dynamic principles of invariance, we are largely on terra incognita. The better understood interaction is the electromagnetic one. In order to describe the interaction of charges with the electromagnetic field, one first introduces new quantities, the so-called electromagnetic potentials, to describe that field. From these, the components of the electromagnetic field can easily be calculated, but not conversely. A further step is then to couple a transformation of the matter field with every transition from a set of potentials to another set of potentials that yields the same electromagnetic field. The combination of these two transformations, one on the electromagnetic potentials, the other on the matter field, is called a *gauge transformation*. Since it leaves the physical situation unchanged, every equation must therefore be invariant.

Unfortunately, our approach can by no means be the same in the case of the other interactions. The strong one, as well as the weak one, includes a group that is, first of all, very much smaller than the gauge group or the group of general coordinate transformations of the general theory of relativity. Instead of the infinity of generators of the gauge and general transformation groups, these interactions have only a finite number of generators. Another difference between the invariance groups of electromagnetism and gravitation, on the one hand, and at least the invariance group of the strong interaction, on the other hand, is that the operations of the former remain valid symmetry operations even if the existence of the other types of interactions is taken into account. The symmetry of the strong interaction, by contrast, is "broken" by the other interactions, i.e., the operations of the group of strong interactions are valid symmetry operations only if the other types of interactions can be disregarded. (On this important subject, see Goldstone et al. [1962], Isham [1984], Witten [1987], and especially Coleman [1985].) In every case the symmetry group helps to determine the interaction operator. But whereas all interactions are invariant under the groups of electromagnetic and gravitational interactions, only the strong interaction is invariant under the group of that interaction.

In studying the particles and the forces between them, one finds that almost all of the macroscopic symmetry principles, which basically represent the symmetries observed by the electromagnetic interactions, hold intact. Some, however, are found to be violated in the weak interactions, and some newly discovered symmetries govern the complex pattern of mutual interaction between the particles. All of these symmetries fall into one of four classes.

1. First, there are the space-time symmetries, which include symmetries resulting from relativistic invariance and symmetries resulting from reflection invariance and time-reversal invariance. These symmetries possess a direct geometrical significance, and they are thoroughly incorporated into the theoretical scheme that describes the elementary particles. Nevertheless, we must consider the possibility of deviations from the space-time continuum structure.

2. In the second class, there is one and only one symmetry, and that is charge-conjugation symmetry, which states roughly that to every particle there is a charge-conjugate particle with the opposite charge. The concept of charge-conjugate symmetry is a purely quantum mechanical concept, and is not related to any geometrical concepts such as rotational invariance. It derives its origin from the Dirac theory of the electron, which in turn is, as viewed today, a logical consequence of the fusion of quantum theory with the requirement of relativistic invariance.

3. In this class, we range four conservation laws: the conservation of the number of heavy particles, the conservation of charge, the conservation of light fermions, and the conservation of isospin[4], which is, however, only approximately conserved. These are conservation laws, and yet one can group them together as representing a type of symmetry principle. Nevertheless, it must be noted that the isospin associated with a conservation law has never been observed, even though it conserves charge, angular momentum, and baryon number. Such a conservation law is forbidden by the fact that is does not conserve isospin.

4. In the final class of symmetries, we group the symmetry between identical particles and a type of symmetry called crossing symmetry that relates reaction rates of the two processes (for example, the relations between the amplitudes for the probability of some collisions), thus

$$A + B \rightarrow C + D$$
$$A + D \rightarrow C + B.$$

According to some physicists, this kind of symmetry forms a transition between the geometrical and dynamical principles.

We notice that as the domain of physics expands, and as symmetry principles increase in number, that increase does seem to follow interdependent and related lines. To understand the relationship and the conceptual unification between them is a major challenge. It seems that three related questions in this pattern of symmetry laws are particularly worth studying. The first is that of the symmetry principle that gives rise to isotopic spin conservation. It is not at all clear whether the isotopic spin conservation is a consequence of an isotopic spin space rotation, or if the isotopic spin space must rather be deduced from the conservation law itself. The second fundamental question relates to the violation of some of the symmetry laws in the weaker interactions. It was found in the early 1960s that violation of parity conservation, which represents a space-time symmetry concept, occurs rather generally among the weak interactions. But this does not explain why weak interactions are less symmetrical than the stronger ones. Finally, if we represent *CP* (charge conjugation plus space inversion) as the operation that corresponds to space inversion, we thus conclude that all interactions satisfy the properties of all space and time symmetries, i.e., proper Lorentz invariance, space inversion, and time-reversal invariance.

Gauge invariance is the very model of dynamical symmetries, in contrast to Lorentz invariance, which is a geometrical symmetry. As we had already seen, geometrical symmetries are regularities of the laws of motion, but are formulated in terms of physical events; the application of the symmetry transformation yields a different physical situation. By contrast, gauge symmetries are formulated only in terms of the laws of nature; the application of the symmetry transformation merely changes our description of the same physical situation. To be sure, gauge invariance is a symmetry of our description of nature, yet it underlies the dynamics. The first example of this invariance was general relativity, in which the symmetry (the equivalence principle) was used to determine the laws of gravity, or of space-time. Furthermore, the realization that gauge symmetry is based on the fiber bundle (a sophisticated geometrical concept) has provided a deep and beautiful geometrical foundation for gauge symmetry. These facts offer a hint that all fundamental symmetries are local gauge symmetries. Global symmetries are either all broken or approximate or they are the remnants of spontaneously broken local symmetries. Thus, Poincaré invariance can be regarded as the residual symmetry of the Minkowski vacuum under changes of the coordinates.

It seems, then, that the secret of nature is symmetry, but much of the texture of the world is due to mechanisms of symmetry breaking. The phenomenon of spontaneous symmetry breaking of global and local gauge symmetries is special to theories with an infinite number of degrees of freedom. In these degrees of freedom, global symmetries may be realized in two different ways. The first way (sometimes referred to as the Wigner-Weyl model) is standard: the laws of physics are invariant, and the ground state of the theory, the vacuum, is unique and symmetric. This is always the case for quantum-mechanical systems. A second way, however, sometimes called the Nambu-Goldstone mode, in which the ground state is asymmetric, is possible. Such spontaneous symmetry breaking is responsible for magnetism, superconductivity, the structure of the unified electroweak theory, and more.

The search for new kinds of symmetry, which should explain some of the most mysterious features of nature, led recently to the consideration of supersymmetry, a profound and beautiful extension of the geometric symmetries of space-time to include symmetries generated by fermionic (anticommuting) charges. Supersymmetric theories (and especially string theory) have the potential to unify bosons (like photons) and fermions (like electrons) into a single pattern, thus to unify matter and force and to help explain the mysterious fact— the hierarchy problem—that the mass scale of atomic and nuclear physics is so much smaller than the scale determined by gravity.

Chapter 12

Quantum Electrodynamics and Gauge Theory

In this chapter I want to emphasize some facts about quantum electrodynamics, that is, the theory that results from combining electron matter fields with electromagnetic fields—as first formulated in the 1930s by P. Dirac and essentially completed by about 1969 by S. Tomonaga (1968), J. Schwinger (1968), R. P. Feynman (1967), and F. J. Dyson (1949, 1969). We recall first that the theory is based on a local gauge symmetry. Another theory, Einstein's general theory of relativity, is based on a local gauge symmetry, which pertains not to a field distributed through space and time but to the structure of space-time itself. Indeed, every point in space-time can be labeled by four numbers, which give its position in the three spatial dimensions and its sequence in the single time dimension. These numbers are the coordinates of the event, and the procedure for assigning such numbers to each point in space-time is a coordinate system, and the choice of such a coordinate system is clearly a matter of convention. The freedom to move the origin of a coordinate system constitutes a symmetry of nature. Actually, there are three related symmetries: all laws of nature remain invariant when the coordinate system is transformed by translation, by rotation, or by mirror reflection. It is important to note, however, that those symmetries are global in nature. Each symmetry transformation can be defined as a formula for finding the new coordinates of a point vis-à-vis its old coordinate. These formulas must be applied simultaneously, and in the same way, to all the points.

In quantum electrodynamics, the symmetry operation consists of a local phase change in the electron field; each such phase shift is accompanied by an interaction with the electromagnetic field. Imagine an electron undergoing two consecutive phase shifts: the emission of a photon and then the absorption of the photon. If the sequence of the phase shifts were reversed, the final result would be the same. It follows that an unlimited series of phase shifts can be made, and the final result will be simply their algebraic sum, no matter what their sequence. On the contrary, in the Yang-Mills theory, where the symmetry operation is a local rotation of the isospin arrow, the result of multiple transformations may be rather different. Recall that the Yang-Mills theory is applicable to weak isospin[1] and therefore to electroweak unified theory $SU(2) \times U(1)$.

Suppose a hadron[2] is subjected to a gauge transformation, A, followed soon after by a second transformation, B; at the end of this sequence the isotopic-spin arrow is found in the orientation that corresponds to a proton. Now suppose that the same transformation is applied to the same hadron but in the reverse sequence: B followed by A. In general, the final state will not be the same; the particle may be a proton instead of a neutron. The net effect of the two transformations depends explicitly on the sequence in which they are applied. Because of this distinction, quantum electrodynamics is called, mathematically, an Abelian theory, with the symmetry group $U(1)$ (the gauge field, which mediates the interaction between the charged 1/2-spin fields, is the electromagnetic field) and the Yang-Mills theory is called a non-Abelian one, with the local (gauge) symmetry group $SU(2)$ for electroweak interaction and the local (gauge) symmetry group $SU(3)$ for strong interaction. Abelian groups are made up of transformations that, when applied one after another, have the commutative property; non-Abelian groups are not commutative. (The terms are borrowed from the mathematical theory of groups created by the Norwegian mathematician N. H. Abel.) Like the Yang-Mills theory, the general theory of relativity is non-Abelian. Even the larger theory in which the electromagnetic interaction has been incorporated is non-Abelian. For now, at least, it seems that all the forces of nature are governed by non-Abelian gauge theories.

This important and surprising result (i.e., the asymmetry of certain fundamental laws of physics) spurred a vast new investigation, still active today, into spontaneous symmetry-breaking. The central question now seems to be the connection between the symmetry-breaking that occurs in the behavior of certain elementary particles at a certain level of size and temperature, and the geometrical structure of space at that same level. More precisely, it has been hypothesized that a symmetry-breaking occurs when there is a change (or degeneration) in the space structure, or, mathematically speaking, a jump from a group to a "poorer" group of the field or the interaction concerned. Nothing, however, prevents us from believing that if there is a richer group containing the two others as subgroups, the difficulty may thereby be removed (see below for further considerations on this point).

Mathematically, the phenomenon of "symmetry breaking" can be explained as follows. Suppose we are presented with a vector bundle V with structure group G. It might happen that under some conditions the structure group of V can be reduced to a subgroup G_0. This phenomenon of gauge symmetry breaking plays a central role in particle physics—more precisely, in the Weinberg-Salam-Glashow model of weak interactions. Suppose that at some low mass scale m, the gauge group G is effectively reduced to a subgroup G_0. Even if the representations R_1 and R_2 are inequivalent as representations of G, they may be equivalent as representations of G_0. In this case, the fermions that were kept massless by the inequivalence of R_1 and R_2 will be able to gain masses of order m. This is precisely what seems to happen in nature. At a mass scale of order $10^{-17} M_{pl}$, the gauge group $SU(3) \times SU(2) \times U(1)$ is reduced to $SU(3) \times U(1)$. At this point, some of the gauge fields become massive. At the same time, because the representations R_1 and R_2 are isomorphic as representations of $SU(3) \times U(1)$, the light fermions can and do gain mass. Many facts of this symmetry-breaking process are not yet understood. For example, why is the mass scale associated with symmetry breaking so tiny compared to the natural mass scale M_{pl}? It is pretty clear, however, that the idealization in which the masses of the particles are all zero is the situation in which the gauge group $SU(3) \times SU(2) \times U(1)$ is not broken to a subgroup.

Consider further the basic decomposition $S = S_+ \otimes S_-$ of the spinor representation S into spinors $S_\pm$ of positive and negative chirality. The distinction between S_+ and S_- here is a matter of convention. Under a change of orientation of space-time, called by physicists a *parity transformation*, S_+ and S_- are exchanged. The representations R_1 and R_2 are therefore exchanged by parity. If we assume that the laws of nature are invariant under parity, then R_1 and R_2 must be isomorphic. The explanation of the lightness of the fermions therefore rests on parity violation. In the 1950s, however, it was discovered that it is the weak interactions that violate parity. Conversely, parity is conserved by strong and electromagnetic interactions; this is the statement that R_1 and R_2 are isomorphic as representations of $SU(3) \times U(1)$. In hopes of overcoming this contradiction, physicists have lately contemplated the possibility of extending the observed gauge group G to a larger group G, as $SU(5)$, which contains $SU(3) \times SU(2) \times U(1)$, or $SO(10)$, or the exceptional group E_6.

In physical terms, however, the problem can be expressed in quite a different manner. It is well-known that one of the serious difficulties of the Yang-Mills theory is that where isospin symmetry is becoming exact, the result is that protons and neutrons are indistinguishable, a situation that is obviously contradictory.

Even more troubling is the prediction of electrically charged photons. Because it must have an infinite range, the photon is necessarily massless. Imposing a mass on the quanta of the charged fields does not make the fields disappear, but it does confine them to a finite range. If the mass is great enough, the range can be made as small as one wishes. As the long-range effects are removed, the existence of the fields can be reconciled with experimental observations. The modified Yang-Mills theory is then easier to understand, but the theory still must be given a quantum-mechanical interpretation.

The problem of infinities turned out to be more severe than it had been in quantum electrodynamics, and the standard recipe for renormalization would not solve the difficulty. In this respect, the fundamental idea of the Higgs mechanism was to include in the modified Yang-Mills gauge theory an extra field, one having the peculiar property that it does not vanish in the vacuum. One usually thinks of a vacuum as a space containing nothing, but in physics that vacuum is defined more precisely as the state in which all fields have their lowest possible energy. For most fields, the energy is minimized when the value of the field is zero everywhere, or in other words when the field is "turned off." An electron field, for example, has its minimum energy when there are no electrons. The effect of the Higgs field is to provide a frame of reference in which the orientation of the isotopic-spin arrow can be determined.

The Higgs field can be represented as an arrow superposed on the other isotopic-spin indicators in the imaginary internal space of a hadron. What distinguishes the arrow of the Higgs field is that it has a fixed length, established by the vacuum value of the field. The orientation of the other isotopic-spin arrows can then be measured with respect to the axis defined by the Higgs field. In this way, a proton can be distinguished from a neutron. It might seem that the introduction of the Higgs field would spoil the gauge symmetry of the theory, and thereby lead again to insoluble infinities. In fact, however, the gauge symmetry is not destroyed but merely canceled. The symmetry specifies that all the laws of physics must remain invariant when the isotopic-spin arrow is rotated in an arbitrary way from place to place. This implies that the absolute orientation of the arrow cannot be determined, since any experiment for measuring the orientation would have to detect some variation in a physical quantity when the arrow is rotated. With the inclusion of the Higgs field, the absolute orientation of the arrow still cannot be determined, because the arrow representing the Higgs field also rotates during a gauge transformation. All that can be measured is the angle between the arrow of the Higgs field and the other isotopic-spin arrows, or in other words their relative orientation.

The Higgs mechanism is an example of the process called spontaneous symmetry breaking, which was already well established in other areas of physics. The concept was first put forward by W. Heisenberg in his description of ferromagnetic materials. Heisenberg pointed out that the theory describing a ferromagnet has perfect geometric symmetry, in the sense that it gives nonspecial distinction to any one direction in space. When the material becomes magnetized, however, there is one axis—the direction of magnetization—that can be distinguished from all other axes. The theory is symmetrical, but the object it describes is not. Similarly, the Yang-Mills theory retains its gauge symmetry with respect to rotations of the isotopic-spin arrow, but the objects described— protons and neutrons—do not express the symmetry. Philosophically, that fact led us to make the distinction between the "ontological" or "objectal" level and the "operational" or "theoretical" level of physical entities; moreover, the first level cannot be reduced to the latter one.

Despite all these difficulties, the Yang-Mills theory had begun life as a model of the strong interactions, but by the time it had been renormalized, interest in it centered on applications to weak interactions. In 1967, S. Weinberg, A. Salam, and C. Ward proposed a model of the weak interactions, based on a version of the Yang-Mills theory, in which the gauge quanta take on mass through the Higgs mechanism. The Weinberg-Salam-Ward model actually embraces both the weak force and electromagnetism. The conjecture on which the model is ultimately founded is a postulate of local invariance with respect to isotopic spin; in order to preserve that invariance, four photon-like fields are introduced, rather than the three of the original Yang-Mills theory. The fourth photon, which could be identified with some primordial form of electromagnetism, corresponds to a separate force, which had to be added to the theory without explanation. For this reason, the model should not be called a unified field theory.

If one were to search for a nonlinear generalization of Maxwell's equation, with which to explain elementary particles, there are various symmetry properties one would require. These are:

1. *external symmetries* under the Lorentz and Poincaré groups and under the conformal group, if one is taking the rest-mass to be zero,
2. *internal symmetries* under groups like $SU(2)$ or $SU(3)$ (these are approximate symmetries)[3] to account for the known features of elementary particles, and
3. *covariance*, or the ability to be coupled to gravitation by considering a curved space-time.

Gauge theories satisfy these basic requirements, because they are geometric in character. In fact, on the mathematical side, gauge theory is a well-established branch of differential geometry known as the theory of fiber bundles with connection. It has much in common with Riemannian geometry, which provided Einstein with the basis for his theory of general relativity. If the current expectations of Yang-Mills theory are eventually fulfilled, it will in some measure justify Einstein's view that the basic laws of physics should all be in geometrical form.

Chapter 13

Vacuum as the Source
of Asymmetry

In this chapter we examine the relationship between the properties of the quantum vacuum and the breaking-symmetry phenomenon. In particular, we study the mechanism responsible for breaking the electroweak gauge symmetry, giving mass to the W and Z bosons. Is there a connection between these broken symmetries and the quantum vacuum low-energy of scalar particles and fields like Higgs fields as predicted by the Goldstone theorem?[1]

WHAT IS VACUUM?

In the nineteenth century, in order to understand how the electromagnetic force, and later the electromagnetic wave, could be transmitted in space, the vacuum came to be viewed as a medium called *æther*. In his note *Experimental Researches in Electricity* (1839), Faraday wrote:

> For my own part, considering the relation of a vacuum in the magnetic force and the general character of magnetic phenomena external to the magnet, I am more inclined to the notion that in the transmission of the force there is such an action, external to the magnet, than that the effects are merely attraction and repulsion at a distance. Such an action may be a function of the aether; for it is not at all unlikely that, if there be an aether, it should have other uses than simply the conveyance of radiations.

But since at that time the nonrelativistic Newtonian mechanics was the only one available, the vacuum was thought to provide an absolute frame that could be distinguished from other moving frames by measuring the velocity of light. As is well-known, this led to the downfall of æther and the rise of relativity. We know now that vacuum is Lorentz-invariant, which means that just by our running around and changing the reference system we are not going to alter the vacuum. But Lorentz invariance does not embody all physical characteristics. We may still ask: What is this vacuum state?

In the modern treatment, we define the vacuum as the lowest-energy state of the system. It has zero 4-momentum. In most quantum field theories, the vacuum is used only to enable us to perform the mathematical construct of a Hilbert space. From the vacuum state, we build the one-particle state, then the two-particle state, and so forth. The resulting Hilbert space, we hope, will eventually resemble our universe. From this approach, different vacuum state means different Hilbert space, and therefore different universe.

From Dirac's hole theory we know that the vacuum, although Lorentz-invariant, is actually quite complicated. In general, we may expect the vacuum to be as complex as any spin-0 field $\phi(x)$ at the zero 4-momentum limit: vacuum $\sim \phi$ at 4-momentum $k_\mu = 0$. It is conceivable that, like a spin-0 field, the vacuum state may carry quantum numbers such as isospin $\vec{I}$, parity P, strangeness S, etc. In this context, we may ask: could the vacuum be regarded as a physical medium? If under suitable conditions the properties of the vacuum, like those of any medium, can be altered physically, then the answer could be affirmative. Otherwise, it might degenerate into semantics. The analysis given below will be based primarily on two of the most remarkable phenomena in modern physics: (i) missing symmetry, and (ii) quark confinement. We will begin with the first.

MISSING SYMMETRY

If we add up the symmetry quantum numbers such as $\vec{I}$, S, P, C, CP, ..., of all matter, we find these numbers to be constantly changing:

$$(13.1) \qquad d/dt \left\{ \begin{array}{l} \vec{I} \\ S \\ P \\ C \\ CP \\ \cdot \\ \cdot \\ \cdot \end{array} \right\}_{\text{matter}} \neq 0.$$

Aesthetically, this may appear disturbing. Why should nature abandon perfect symmetry? Physically, this also seems mysterious. What happens to these missing quantum numbers? Where do they go? Can it be that matter alone does not form a closed system? If we also include the vacuum, then perhaps symmetry may be restored:

$$(13.2) \qquad d/dt \left\{ \begin{matrix} \vec{I} \\ S \\ P \\ C \\ CP \\ \cdot \\ \cdot \\ \cdot \end{matrix} \right\}_{\text{matter+vaccum}} = 0.$$

As a bookkeeping device, this is clearly possible. It also forms the basic idea underlying the important topic of spontaneous symmetry breaking, developed by Y. Nambu and J. Goldstone. In such a scheme, one often assumes that there exists some phenomenological spin-0 field ϕ that can carry the missing quantum number, and whose vacuum expectation value is not zero:

$$(13.3) \qquad \phi_{\text{vac}} \equiv \langle \text{vac} \,|\phi|\, \text{vac} \rangle \neq 0.$$

Consequently, the observed asymmetry can be attributed entirely to the state vector of our universe, not to the physical law. Conversely, unless we have other links connecting matter with vacuum, how can we be sure that this idea is valid, and not merely a tautology?

A way out of this dilemma is to realize that in vacuum ~ ϕ at 4-momentum $k_\mu = 0$, the restriction $k_\mu = 0$ for the vacuum state is only a mathematical idealization. After all, the universe very likely does have a finite radius, and k_μ is therefore never strictly zero. So far as the microscopic system of particle physics is concerned, there is a little difference between $k_\mu = 0$ and k_μ nearly 0; the latter corresponds to a state that varies only very slowly over a large space-time extension. This means that if the idea expressed by $\phi_{\text{vac}} \equiv \langle \text{vac} \,|\phi|\, \text{vac} \rangle \neq 0$ is correct, then under suitable conditions, we must be able to produce excitations, or domain structures, in the vacuum. In such an excited state, there exists a volume Ω whose size is >> the relevant microscopic dimension; inside Ω we have the expectation value $\langle\phi(x)\rangle \neq \phi_{vac}$, but outside Ω $\langle\phi(x)\rangle = \phi_{vac}$. The symmetry properties inside Ω can then differ from those outside.

VACUUM EXCITATION

How can we produce such a change in $\langle\phi(x)\rangle$? The problem is analogous to the formation of domain structures in a ferromagnet. We may draw the analogue:

$$\langle\phi(x)\rangle \leftrightarrow \text{magnetic spin,}$$
$$J = \text{matter source} \leftrightarrow \text{magnetic field}$$

as shown in fig. 21.

In the case of a very large ferromagnet, whose spins interact linearly with the magnetic field, we can create a domain structure by applying an external magnetic field over a large volume. Furthermore, once the domain is created, we can remove the external field. Depending on the long-range forces, the surface energy, and other factors, such a domain structure may persist even after the external magnetic field is removed. Similarly, by applying over a large volume any matter source J that has a linear interaction with $\phi(x)$, we may hope to create a domain structure in $\langle\phi(x)\rangle$. Depending on the dynamical theory, such domains may also remain as physical realities, even after the matter source J has been removed.

As an illustration, we may consider a local scalar field theory. The Lagrangian density is

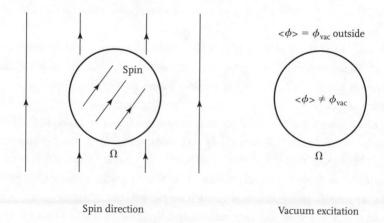

Spin direction   Vacuum excitation

Fig. 21. Domain structures in a ferromagnet vs. those in the vacuum.

(13.4) $$\mathcal{L}_\phi = -1/2(\partial\phi/\partial\phi_\mu) - U(\phi),$$

where the absolute minimum of U is at $\phi = \phi_{vac}$ with $U(\phi_{vac}) = 0$. Since we are interested in the long-wavelength limit of the field, the scalar field $\phi(x)$ is used only as a phenomenological description. Let us now introduce an external source $J(x)$. The simplest example is to assume J to be constant inside a large volume Ω, but zero outside. For a sufficiently large Ω, we may neglect the surface energy. The energy of the system then becomes

(13.5) $$[U(\phi) + J\phi]\Omega.$$

Its minimum determines the expectation value of ϕ inside Ω. The graph in fig. 22 illustrates how the new expectation value $\bar\phi = \langle\phi(x)\rangle$ can be changed under the influence of J.

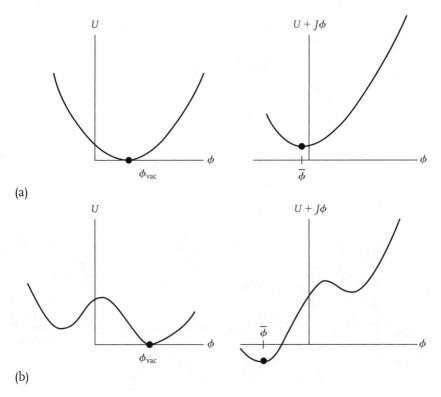

Fig. 22. Change of $\bar\phi = \langle\phi(x)\rangle$ due to a constant external matter source J. In case (a), $\bar\phi$ changes continuously with J. In case (b), as J increases, there is a critical value at which $\bar\phi$ makes a sudden leap.

If the missing symmetry is due to $\phi_{\mathrm{vac}} \neq 0$, then by changing $\bar{\phi}$ we may alter the symmetry properties inside Ω dynamically.

CP NONCONSERVATION AND SPONTANEOUS SYMMETRY BREAKING

We discuss now one of the simplest examples illustrating the phenomenon of spontaneous symmetry breaking. Our purpose is to give a theory in which (i) the Lagrangian is invariant under CP and T, but (ii) its S-matrix violates CP and T symmetry.

Let us assume the system consists of a spin $-1/2$ Dirac field ψ and a spin -0 Hermitian field ϕ. The Lagrangian density is

$$(13.6) \quad \mathscr{L} = -1/2 (\partial \phi / \partial x_\mu)^2 - U(\phi) - \psi^\dagger \gamma_4 (\gamma_\mu \partial / \partial x_\mu + m) \, \psi - ig \psi^\dagger \gamma_4 \gamma_5 \, \psi \, \phi,$$

where

$$(13.7) \qquad\qquad U(\phi) = \frac{1}{8} \kappa^2 (\phi^2 - \rho^2)^2.$$

See fig. 23. From Hermiticity[2], the parameters m, g, ρ, and κ must be real. It can then be readily verified that $\mathscr{L}$ is invariant under T, C, and P, where

$$(13.8) \qquad\qquad T\phi(\vec{r},t)T^{-1} = -\phi(\vec{r},-t),$$

$$(13.9) \qquad\qquad C\phi(\vec{r},t)C^{-1} = \phi(\vec{r},t),$$

and

$$(13.10) \qquad\qquad P\phi(\vec{r},t)P^{-1} = -\phi(-\vec{r},t).$$

The corresponding transformations of ψ are given by

$$(13.11) \qquad\qquad C\psi(x)C^\dagger = \eta_C \psi^C(x),$$

$$(13.12) \qquad\qquad P\psi(\vec{r},t)P^\dagger = \eta_P \gamma_4 \, \psi(-\vec{r},t),$$

and

(13.13)
$$T\psi(\vec{r},t)T^{-1} = \eta_t\sigma_2\,\psi(-\vec{r},t).$$

Because U is a fourth-order polynomial in ϕ, the theory is renormalizable.

The vacuum expectation value of ϕ is determined by the minimum of $U(\phi)$. We have either $\langle\phi\rangle_{vac} = \rho > 0$ or $\langle\phi\rangle_{vac} = -\rho$. In either case, since ϕ is of $T = -1$, $CP = -1$, and $P = -1$, a nonzero expectation value of ϕ implies that the vacuum state is not an eigenstate of T (nor of CP and P). The T-symmetry of the Lagrangian requires that if $\langle\phi\rangle_{vac} = \rho$ is a solution, then $\langle\phi\rangle_{vac} = -\rho$ must also be one. These two solutions transform into each other under T, but by itself neither is invariant under T. (It is also not invariant under CP and P, though it is under C and CPT.)

Because of quantum effects, the field ϕ fluctuates around its vacuum expectation value. We may choose $\langle\phi\rangle_{vac} = \rho$, and write $\phi = \rho + \delta\phi$. In terms of $\delta\phi$, the potential U becomes

(13.14)
$$U = \frac{1}{2}\mu^2(\delta\phi)^2 + \frac{1}{2}\kappa^2\rho(\delta\phi)^3 + \frac{1}{8}\kappa^2(\delta\phi)^4,$$

where $\mu = \kappa\rho$, which is the mass of the fluctuating field $\delta\phi$.

In order to exhibit more clearly the T-violating character of the solution, we may perform a unitary transformation under which ϕ is unchanged, but where

(13.15)
$$\psi \to e^{-i\frac{1}{2}\gamma_5\alpha}\psi.$$

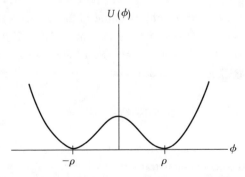

Fig. 23. The potential energy density $U(\phi) = 1/8\,\kappa^2(\phi^2 - \rho^2)^2$.

Therefore, the quadratic expressions

(13.16)
$$\psi^\dagger \gamma_4 \psi \to \psi^\dagger e^{\frac{1}{2}i\gamma_5\alpha} \gamma_4 e^{-\frac{1}{2}i\gamma_5\alpha}\psi = \psi^\dagger \gamma_4 e^{-i\gamma_5\alpha}\psi =$$
$$= \psi^4 \gamma_4 (\cos\alpha - i\gamma_5 \sin\alpha)\psi$$

and

(13.17)
$$i\psi^\dagger \gamma_4 \gamma_5 \psi \to i\psi^\dagger e^{i\frac{1}{2}i\gamma_5\psi} \gamma_4\gamma_5 e^{-i\frac{1}{2}i\gamma_5\alpha}\psi = i\psi^\dagger \gamma_4\gamma_5 e^{-i\gamma_5\alpha}\psi =$$
$$= \psi^\dagger \gamma_4 (\sin\alpha - i\gamma_5 \cos\alpha)\psi.$$

Hence, by choosing

(13.18)
$$\tan\alpha = g\rho/m,$$

we have

(13.19)
$$\psi^\dagger \gamma_4 (m + ig\rho\gamma_5) \psi \to \psi^\dagger \gamma_4 M\psi,$$

where

(13.20)
$$M = (m^2 + g^2 \rho^2)^{1/2}.$$

By substituting (13.17) into (13.20), we find that the Lagrangian density $\mathcal{L}$ becomes

(13.21)
$$-\frac{1}{2}\left(\frac{\partial}{\partial x_\mu}\delta\phi\right)^2$$
$$-U - \psi^\dagger \gamma_4 \left(\gamma^\mu \frac{\partial}{\partial x_\mu} + M\right)\psi - g\psi^\dagger \gamma_4 (\sin\alpha + i\gamma_5 \cos\alpha)\psi\,\delta\phi.$$

Since the operator $\psi^\dagger \gamma_4 \psi$ is of $P = 1$, $C = 1$, and $T = 1$ while the operator $i\psi^\dagger \gamma_4 \gamma_5 \psi$ is of $P = -1$, $C = 1$, and $T = -1$, any exchange of the quantum $\delta\phi$ would give an interference term between these two operators that violates T, P, and CP; but the product symmetry CPT remains intact. Fig. 24 gives an example of such an interference term, whose amplitude is

(13.22)
$$A_- = g^2 \sin\alpha \cos\alpha \, (\kappa^2 + \mu^2)^{-1},$$

where κ denotes the 4-momentum transfer.

The model just discussed illustrates the basic mechanism of a spontaneous T violation. (The same discussion applies to CP and P violation as well.) One as-

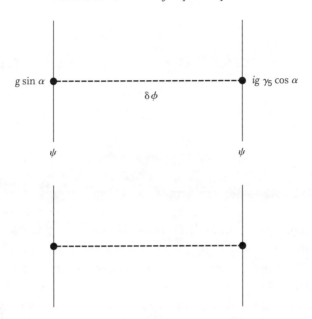

Fig. 24. A *T*-violating scattering diagram illustrating the exchange of the quantum of $\delta\phi$, where $\delta\phi \equiv \phi - \langle\phi\rangle_{\text{vac}}$.

sumes that the ground state of the system (called the vacuum) has nonzero expectation value $\langle\phi\rangle_{\text{vac}}$, where ϕ is a $T = -1$ phenomenological spin-0 field; the vacuum is noninvariant under T, even though the Lagrangian satisfies T invariance.

The T invariance of the Lagrangian implies that the vacuum must have a double degeneracy. It is interesting to examine the barrier penetration between these two degenerate solutions: $\langle\phi\rangle_{\text{vac}} = \rho$ and $-\rho$ of fig. 23. By enclosing the entire system in a finite volume Ω with a periodic boundary condition, we may expand ϕ in terms of the usual Fourier series

(13.23)
$$\phi = q + \sum_{k\neq0} 1/\sqrt{\Omega}\,\phi_k\,e^{i\vec{k}\cdot\vec{r}},$$

where q is independent of $\vec{r}$. Substituting this expression into (13.21), we find

(13.24)
$$L = \int \mathcal{L}\,d^3r = \frac{1}{2}\dot{q}^2\Omega - U(q)\Omega + \cdots$$

where . . . depends on ϕ_k and the fermion field ψ. The barrier-penetration amplitude may be crudely estimated by concentrating on the q–degree of freedom. We may set $\phi_k = 0$ and $\psi = 0$. The Hamiltonian then becomes

(13.25) $H = (1/2\Omega)\, p^2 + (U(q))\,\Omega,$

where $p = \Omega q$. According to the W.K.B. approximation in quantum mechanics, the barrier-penetration amplitude is

(13.26) $\sim \exp\left\{-\Omega \int_{-\rho}^{\rho} [2U(q)]^{1/2}\,dq\right\},$

which goes to zero exponentially as the volume of the system approaches infinity.

From (13.20) we see that the fermion mass changes from m to M when the expectation value of ϕ varies from 0 to $\langle\phi\rangle_{vac} = \rho$. Likewise, the CP-violating amplitude A_- also depends on $\langle\phi\rangle_{vac}$. Hence, if we follow the above discussion on vacuum expectation, by applying a matter source J over a large volume Ω we may alter $\langle\phi\rangle$ inside Ω, and thereby change the mass of the particle and the symmetry-violating amplitude.

The above example demonstrates the essence of the spontaneous symmetry-breaking mechanism. The Lagrangian is invariant under a certain group G of symmetry transformations. But the vacuum state is not, and that gives rise to symmetry-violating phenomena. By applying G onto the vacuum state, we must generate other states that are degenerate with the vacuum. In a realistic cosmological model, the volume Ω of the universe may be expected to be finite. In general, there would be nonzero, but very small, barrier-penetration amplitudes between these different "vacua," which can lift the degeneracy. Although such effects can be safely neglected at our present stage of evolution, they may have been important at a much earlier period when Ω was still of microscopic dimensions.

Chapter 14

Topological Quantum Field Theories and Gauge Theories

A Far-Reaching Interface between Geometry and Physics

It is well known that in quantum mechanics, the position and velocity of a particle are noncommuting operators acting on a Hilbert space, and classical notions such as "the trajectory of a particle" do not apply.

But in nineteenth- and early twentieth-century physics, many aspects of nature were described in terms of fields—the electric and magnetic fields that enter into Maxwell's equations, and the gravitational field governed by Einstein's equations. It was clear that, since fields interact with particles, then to give an internally coherent account of nature, the quantum concepts must be applied to fields as well as particles. When this is done, quantities such as the components of the electric field at different points in space-time become noncommuting operators. When one constructs a Hilbert space in which these operators act, one finds many surprises. For one thing, the distinction between fields and particles breaks down, since the Hilbert space of a quantum field is constructed in terms of particle-like excitations; and for another, conventional particles, such as electrons, are reinterpreted as arising from the quantization of a field, and in the process one finds the prediction of "antimatter": for every particle, there must be a corresponding antiparticle, with the same mass and opposite electric charge; predicated by Dirac on the basis of quantum field theory, the "positron" of oppositely charged antiparticle of the electron was discovered in cosmic rays.

The quantum field theories (QFTs) that have proved to be most important in describing elementary particle physics are gauge theories. As we have already seen, the classical example of a gauge theory is the theory of electromagnetism. The gauge theory is the Abelian group $U(1)$. If A denotes the $U(1)$ gauge connection, which locally can be regarded as a one-form on space-time, then the curvature or electromagnetic field tensor is the two-form $F = dA$, and Maxwell's equations read

$$(14.1) \qquad\qquad 0 = dF = d * F.$$

(Here $*$ is the Hodge duality operator; Hodge indeed introduced his celebrated theory of harmonic forms as a generalization to p-forms of Maxwell's equations.)

Yang-Mills theory, or non-Abelian gauge theory, can, at the classical level, be described similarly, with $U(1)$ replaced by a more general compact gauge group G. The definition of the curvature must be modified to $F = dA + A \wedge A$, and Maxwell's equations are modified to the Yang-Mills equations, $0 = d_A F = d_A * F$, where d_A is the gauge covariant extension of the exterior derivative. These equations can be derived from the Yang-Mills Lagrangian

$$(14.2) \qquad\qquad \mathscr{L} = \frac{1}{4} g^2 \int \mathrm{Tr}\, F \wedge * F,$$

where Tr denotes an invariant quadratic form on the Lie algebra of G. The Yang-Mills equations are nonlinear, in contrast to the Maxwell equations, but like the Einstein equations for the gravitational field they are not explicitly soluble in general. But they do have certain properties in common with the Maxwell equations, and in particular they describe at the classical level massless waves that travel at the speed of the light.

Quantum electrodynamics (QED) gave a very accurate account of quantum behavior of electromagnetic fields and forces. The question arose of whether the non-Abelian analogue was important for describing other forces in nature, notably the weak force and the strong or nuclear force. The masslessness of classical Yang-Mills waves was a serious obstacle to applying Yang-Mills theory to the other forces, because the weak and nuclear forces are not associated with long-range fields or massless particles.

In the 1960s and 1970s, these obstacles to physical applications of non-Abelian gauge theory were overcome. In the case of the weak force, this was accomplished by the Weinberg-Salam-Glashow electroweak theory with gauge group $H = SU(2) \times U(1)$. The masslessness of classical Yang-Mills waves was avoided by elaborating the theory with an additional "Higgs field": this is a scalar field, transforming in a two-dimensional representation of H, whose nonzero and approximately constant value in the vacuum state reduces the structure group from H to a $U(1)$ subgroup (diagonally embedded in $SU(2) \times U(1)$). This theory describes both the electromagnetic and weak forces, in a more or less unified way, because of the reduction of the structure group to $U(1)$.

The solution to the problem of massless Yang-Mills fields for the strong interactions was of a completely different nature. The solution was not obtained by adding additional fields to Yang-Mills theory but by discovering a remarkable property of the quantum Yang-Mills theory itself (i.e., the quantum theory whose Lagrangian has been given above). This property is called "asymptotic freedom." When a quantum theory is asymptotically free, this means, roughly, that the quantum behavior at short distances is very similar to the classical behavior, but that the classical theory is not a good guide to the quantum behavior at long distances. Asymptotic freedom, together with other experimental and theoretical discoveries made in the 1960s and 70s made it possible to describe the nuclear force by a non-Abelian theory in which the group is $G = SU(3)$. The additional fields describe, at the classical level, "quarks," which are spin 1/2 objects somewhat analogous to the electron, but transforming in the fundamental representation of $SU(3)$. The non-Abelian gauge theory of the strong force is called quantum chromodynamics (QCD).

What should now be stressed is that classical non-Abelian gauge theory is very different from the observed world of strong interactions. For QCD to successfully describe the strong force, it must have at the quantum level the following three properties, each of which is dramatically different from the behavior of the classical theory:

1. It must have a "mass gap." That is, there must be some strictly positive constant Δ such that every excitation of the vacuum has energy at least Δ. This point is necessary to explain why the nuclear force is strong but short-ranged.
2. It must have "quark confinement." That is, even though the theory is described in terms of elementary fields, such as the quarks, that transform nontrivially under $SU(3)$, the physical particle states—such as the proton, neutron, and spin—are $SU(3)$-invariant. This point is needed to explain why we never see individual quarks.
3. It must have "chiral symmetry breaking," which means that the vacuum is potentially invariant under a certain subgroup of the full symmetry group that acts on the quark fields. This point is needed to account for the "current algebra" theory of the soft pions that was developed in the 1960s.

A fundamental point to which I would like to shift briefly is the mathematical explanation of the quantum behavior of four-dimensional gauge theory. Whereas classical non-Abelian gauge theory has played an important role in mathematics in the last two decades, especially in the study of three- and four-manifolds, one does not yet have a precise mathematical definition of quantum gauge theory in four dimensions.

Nevertheless, it seems that many mathematical subjects that have been actively studied in the last few decades can be naturally formulated in terms of QFT. New structures spanning analysis, algebra, and geometry have emerged. On the analytical side, to cite just one example, renormalization theory arose from the physics of quantum field theory and statistical physics, and provides the basis for the mathematical investigation of the local (ultraviolet) regularity and the global decay (infrared regularity) of quantum field theories. Surprisingly, the ideas from renormalization theory also apply in other areas of mathematics.

On the algebraic side, investigations of soluble models of quantum field theory and statistical mechanics have led to many new discoveries involving topics such as Yang-Baxter equations, quantum groups, rational conformal field theory, etc.

Geometry abounds with new mathematical structures rooted in quantum field theory. Examples include the Donaldson theory of four-manifolds, the Jones polynomial of knots and its generalizations, mirror symmetry of complex manifolds, elliptic cohomology, $SL(2, \mathbb{Z})$ symmetry in the theory of affine Kac-Moody algebras, and Clifford geometric algebras. (Clifford spaces and structures may shed some light on the origins behind the hidden E_8 symmetry of 11D supergravity and many reveal more important features underlying M, F theory.) QFT has in certain cases suggested new perspectives on mathematical problems. Moreover, QFT must be developed as a mathematical subject that involves, in particular, quantum gauge theory in four dimensions.

A recent fundamental example of the intimate and fruitful interaction between geometry and physics is the development of topological quantum field theories and their relevance to the study of geometry and topology of low-dimensional manifolds. The application of topological quantum theories reflects the enormous interest generated by mathematicians and field theoreticians in building a link between quantum physics through its path integral formulation, on the one hand, and the geometry and topology of low-dimensional manifolds on the other. These are indeed deep links that are only now getting explored. It does appear that the properties of low-dimensional manifolds can be successfully unraveled by relating them to finite-dimensional manifolds of fields.

The idea of a topological field theory was introduced by Witten in 1988 as a rudimentary structure to which, in principle, any quantum field theory reduces at very long distances and low energies. An axiomatic formulation of topological quantum field theories was also proposed by Michael Atiyah in 1988. A number of examples of topological field theories are quite relevant in geometric topology. One of them provides a unified point of view on the knot invariants discovered by V. Jones, and on the associated invariants of 3-manifolds. Another encodes the Donaldson invariants of 4-manifolds and the Floer cohomology groups of 3-manifolds.

Topological quantum field theories are independent of the metric of curved manifold on which they are defined; the expectation value of the energy-momentum tensor is zero, $\langle T_{\mu\nu} \rangle = 0$. These possess no local propagating degrees of freedom; only degrees of freedom are topological. Operators of interest in such a theory are also metric-independent. The mathematics of topological field theories is closely linked with string theory as a theory of gravitation and elementary particles. One possible way to look at it, from the mathematical point of view, is to say that it replaces the finite-dimensional space-time manifolds of conventional quantum field theory by a completely new kind of "stringy" manifold. To define a conventional manifold is to define the commutative algebra of smooth functions on it. A "stringy" manifold is described not by a commutative algebra but by a more sophisticated algebraic structure that is a fairly natural generalization of a commutative algebra.

To illustrate how ideas of quantum field theory can be used to study topology, I shall focus our attention here on recent important developments in Chern-Simons gauge field theory as a topological quantum field theory on a three-manifold. This theory offers a field-theoretic framework for the study of knots and links in a given 3-manifold. The now famous Jones polynomial for knots and links is intimately related to Chern-Simons theory (1974). Witten set up a general field-theoretic framework to study knots and links (1989). Since then, enormous effort has gone into developing an exact and explicit nonperturbative solution of this field theory. The interplay between quantum field theory and knot theory is a fascinating topic, and a promising area of research for both geometry and theoretical physics. Many of the open problems in knot theory have found answers in the process.

Wilson loop operators are the topological operators of the Chern-Simons gauge field theory. Their vacuum expectation values are the topological invariants for knots and links, which do not depend on the exact shape, location, or form of the knot and links, but reflect only their topological properties. The power of this framework is so deep that it allows one to study these invariants not only on simple manifolds such as 3-sphere but also on any arbitrary 3-manifold. The knot and link invariants obtained from these theories are also intimately related to the integrable vertex models in two dimensions. These invariants have also been approached in different mathematical frameworks, and a quantum group approach to these polynomial invariants has been developed. The last two decades or so have seen enormous activity in these directions in algebraic topology. A mathematically important development is that these link invariants provide a method of obtaining a specific topological invariant for 3-manifolds in terms of invariants for framed unoriented links in S^3.

Chern-Simons theory has also played a major role in quantum gravity. Three-dimensional gravity with a negative cosmological constant, itself a topological field theory, can be described by two copies of $SU(2)$ Chern-Simons theory. Even in 4-dimensional gravity, Chern-Simons theories find application. For example, the boundary degrees of freedom of a black hole in four dimensions are described by an $SU(2)$ Chern-Simons field theory. This has allowed an exact calculation of quantum entropy of a nonrotating black hole. The formula so obtained for a Schwarzschild black hole, although agreeing with the Bekenstein-Hawking formula for large areas, goes beyond the semiclassical result.

Chapter 15

Remarks on Kaluza-Klein Theory
and Supergravity

In the 1920s there were the very interesting attempts by Theodor Kaluza, then by Oskar Klein, to unify the relativistic theory of gravitation with Maxwell's theory by introducing a new geometrical framework within which electromagnetism could be coupled with gravity (at least theoretically). The Kaluza-Klein theories, purely geometrical in character, have been worked out in order to encompass two apparently inconsistent physical theories within a unitary theoretical explanation. Actually, even before Einstein's general relativity appeared, the physicist Gunnar Nordström, in 1914, proceeded to unify his theory of gravitation (in which gravity was described by a scalar field coupled to the trace of the energy momentum tensor) with Maxwell's theory, in a most imaginative way. Inspired by Minkowski's four-dimensional space-time continuum, Nordström added yet another space dimension, thus obtaining a flat five-dimensional world. There he introduced an Abelian five-vector gauge field for which he wrote down the Maxwell equations, including a conserved five-current. He then identified the fifth component of the five-vector potential with scalar gravity, whereas the first four components of the five-vector potential he identified with the Maxwell four-potential. With these interpretations in hand, he then noticed that in the cylindrical case (when all dynamical variables become independent of the fifth coordinate) the equations of his five-dimensional "Maxwell"-theory reduced to those of the four-dimensional Maxwell-Nordström electromagnetic-gravitational theory. It is then fair to say that higher-dimensional unification starts with Nordström, who assumed scalar gravity in our four-dimensional world to be a remnant of an Abelian gauge theory in a five-dimensional flat space-time.

The next step was taken by the mathematician Theodor Kaluza in 1919, in the wake of Einstein's general relativity. Kaluza proposed that one may pass to an

Einstein-type theory of gravity in five dimensions, from which ordinary four-dimensional Einstein gravity and Maxwell electromagnetism are to be obtained, upon imposing a cylindrical constraint. More precisely, what this amounts to is starting with a five-dimensional manifold M, which is the product of $M = M^4 \times S^1$ of a four-dimensional space-time M^4 with a circle S^1. The metric $\gamma_{mn}(x, y)$ on the five-manifold M ($m, n = 0, 1, 2, 3, 5$) is a function of both the coordinates x^μ ($\mu = 0, 1, 2, 3$) on M^4 and of $y \equiv x^5$, the coordinate of the circle S^1. It is convenient to replace the fifteen field variables $\gamma_{mn}(=\gamma_{nm})$ by fifteen new field variables $g_{\mu\nu} = g_{\nu\mu}$, A_μ, ϕ, according to the field redefinitions

(15.1)
$$\gamma_{\mu\nu} = g_{\mu\nu} + e^2 \kappa^2 \phi A_\mu A_\nu,$$

$$\gamma_{\mu 5} = \gamma_{5\mu} = e\kappa\phi A_\mu,$$

and

$$\gamma_{55} = \phi.$$

All field quantities, old and new, are periodic functions of the coordinate y on the circle. If $y = \rho\theta$, where θ is the usual angular coordinate and ρ is the radius of the circle, then the period is $2\pi\rho$. Thus any field quantity $F(x, y)$ (F being any of the $g_{\mu\nu}$'s, A_μ's, ϕ, or γ_{mn}'s) admits a Fourier expansion

(15.2)
$$F(x, y) = \sum_{n=-\infty}^{+\infty} F^{(n)}(x) e^{iny/\rho}.$$

Kaluza assumed the five-dimensional dynamics to be governed by a gravitational Einstein-Hilbert action

(15.3)
$$I_5 = -\frac{1}{16\pi G_5} \int \sqrt{|\gamma_5|}\, R_5 d^5 x,$$

where γ_5 is $\det(\gamma_{mn})$, R_5 is the five-dimensional curvature scalar, and G_5 is a five-dimensional counterpart of the gravitational constant. Using the Fourier expansions, the y-dependence becomes explicit, and the y-integration can then be carried out. A four-dimensional action involving an infinity of fields—the Fourier components $A_\mu^{(n)}$, $g_{\mu\nu}^{(n)}(x)$, and $\phi^{(n)}$—then emerges. At this point, Kaluza imposed a "cylindricity" condition: he truncated the action by dropping all harmonics with $n \neq 0$, retaining only the zero modes:

(15.4) $g_{\mu\nu}(x, y) = g_{\mu\nu}^{(0)}(x)$; $A_\mu(x, y) = A_\mu^{(0)}(x)$; $\phi(x, y) = \phi^{(0)}(x)$.

The five-dimensional line element then takes the form

(15.5) $ds_5^2 \equiv \gamma_{mn} dx^m dx^n = ds_4^2 + \phi^{(0)}(x)(dx^5 + e\kappa A_\mu^{(0)}(x) dx^\mu)^2,$

where

(15.6) $ds_4^2 \equiv g_{\mu\nu}^{(0)}(x) dx^\mu dx^\nu$

is the four-dimensional line element corresponding to the metric $g_{\mu\nu}^{(0)}(x)$. The line element (15.6) is invariant under the transformations

$$x_5^\mu \rightarrow x_5^\mu,$$
$$x \rightarrow x + e\kappa\alpha(x^p),$$
(15.7) $$A_\mu^{(0)} \rightarrow A_\mu^{(0)} - \partial_\mu\alpha(x^p),$$
$$\phi^{(0)} \rightarrow \phi^{(0)}, \text{ and}$$
$$g_{\mu\nu}^{(0)} \rightarrow g_{\mu\nu}^{(0)},$$

which we recognize as Abelian gauge transformation à la Weyl. Here, these transformations assume a geometrical meaning as shifts in the fifth coordinate by an amount $\alpha(x^p)$, which depends on ordinary four-space-time. The Abelian gauge symmetry in four dimensions originates in the isometries of the small circle in the fifth dimension.

When the y integration is carried out with the cylindric truncation enforced, the action (15.3), invariant under five-dimensional general coordinate transformation, reduces to a four-dimensional action invariant under both four-dimensional general coordinate transformations and Abelian gauge transformations. This four-dimensional action is, up to a surface term,

(15.8) $I_4 = \int \sqrt{|g_4^{(0)}|} \sqrt{|\phi^{(0)}|} \left[-\frac{1}{16\pi G} R_4^{(0)} + \left(\frac{e^2\kappa^2}{16\pi G} \right) \phi^{(0)} g^{(0)\mu\rho} g^{(0)\nu\sigma} F_{\mu\nu}^{(0)} F_{\rho\sigma}^{(0)} \right],$

with

(15.9) $G = (G_5 / 2\pi\rho), g_4^{(0)} = \det(g_{\mu\nu}^{(0)}), F_{\mu\nu}^{(0)} = \partial_\mu A_\nu^{(0)} - \partial_\nu A_\mu^{(0)},$

and $R_4^{(0)}$ = scalar curvature calculated from the four-metric $g_{\mu\nu}^{(0)}$; our metric convention calls for a minus (plus) sign for a time (space-like dimension).

This action involves a graviton $g_{\mu\nu}^{(0)}$, an Abelian gauge boson $A_\mu^{(0)}$ and a scalar field $\phi^{(0)}$. Kaluza arbitrarily set $\phi^{(0)}$ as constant, in which case I_4 turns into the four-dimensional Einstein-Maxwell action. To be sure, one must have $\phi^{(0)} > 0$ in order to realize the proper relative sign of the Einstein and Maxwell terms, so that energy is positive. This in turn means that the fifth dimension must be space-like; in fact, the extra dimensions must all be space-like. In addition to the invariances under general coordinate transformations and gauge transformations, the action (15.8) also exhibits an invariance under global scale transformations:

(15.10)
$$g_{\mu\nu}^{(0)} \to \lambda^{-1} g_{\mu\nu}^{(0)}$$
$$A_{\mu}^{(0)} \to \lambda^{-3/2} A_{\mu}^{(0)}$$
$$\phi^{(0)} \to \lambda^2 \phi^{(0)}.$$

The field equations of the original five-dimensional theory afford a solution in which the five-dimensional space-time is the direct product of a circle with flat four-dimensional Minkowskian space-time. Then

(15.11)
$$g_{\mu\nu} = \eta_{\mu\nu}, \, A_{\mu} = 0, \, \phi = 1,$$

where $\eta_{\mu\nu}$ is the four-dimensional Minkowski metric $(-1, +1, +1, +1)$. This solution serves as a natural vacuum, and spontaneously breaks the scale invariance (15.10). The massless $\phi^{(0)}$-field is the Nambu-Goldstone boson associated with this spontaneous symmetry breaking. So the zero-mode spectrum includes spin 2 and spin 1 gauge fields and a spin 0 Nambu-Goldstone boson. In the full quantum theory, the spin zero boson is expected to acquire a mass. Of course, the full classical theory contains not only the zero modes, but also the $n \neq 0$ harmonics of eq. (15.2). The action (15.3) determines their spins, masses, and couplings. They all have spin ≤ 2, and they all are massive. The nth harmonics have mass

(15.12)
$$m_n = \frac{|n|}{\rho},$$

where ρ, as before, is the radius of the small circle in the fifth dimension. The couplings of these harmonics to the gauge field $A_{\mu}^{(0)}$ are determined from the action (15.3), and these harmonics do carry electric charge

(15.13)
$$g_n = n \frac{4\sqrt{\pi G}}{\rho}.$$

Remarkably, electric charge is quantized, because the fifth dimension is compact. We see that the elementary charge is

(15.14) $e: \dfrac{4\sqrt{\pi G}}{\rho}$

and the corresponding fine structure constant is

(15.15) $\alpha = 4G/\rho^2.$

If α is to correspond to the $U(1)$ subgroup of the grand-unification group, then $\alpha \sim 1/100$, so that the circumference of the small circle $l \equiv 2\pi\rho \sim 100 \sqrt{G} \sim 10^{-17} GeV^{-1}$. The circle must be very small indeed; a size of about 100 Planck lengths could hardly have been detected as yet. Nevertheless, this is large enough to call into question grand-unification in four-dimension: the scales at which the grand-unification group is to reveal itself unbroken are close to the scales at which the extra dimensions would become manifest. Of course, to make all this applicable in a world with strong and electroweak interactions, one has to introduce more than one extra dimension.

The Kaluza work was unknown until 1926, when Oskar Klein rediscovered Kaluza's theory. (Einstein delayed the publication of Kaluza's paper for two years.) Klein noted the quantization of the electric charge and hoped Kaluza theory would underlie quantum mechanics (see chapter 8). The relativistic generalization of Schrödinger's equation was carried out independently by many authors: Schrödinger, Klein, Gordon, Fock, and others. This equation, now commonly known as the Klein-Gordon equation, was arrived at by both Klein and Fock, starting from Kaluza theory: a zero mass-wave equation in five dimensions yields four-dimensional Klein-Gordon equations for the individual harmonics. It must be noted that this early work has been viewed as a mathematical trick devoid of any physical significance. Nevertheless, this mathematical idea will be very fruitful in revealing further developments of the theory, especially in supergravity and string theories. Oskar Klein, who comes closest to the modern point of view, discussed the higher harmonics and the size of the small circle in 1926. Later, Einstein and Bergman also adopted such a point of view. A purely mathematical approach (a projective interpretation of the fifth coordinate) was developed by O. Veblen, W. Pauli, C. Jordan, and others. Jordan appears to have been the first to realize the importance of including the scalar field $\phi^{(0)}$ in the new five-dimensional theory.

Remarkably, the most recent work on superstrings incorporates both the ideas of Nordström and the subsequent ideas of Kaluza and Klein. But there was no real reason to extend the Kaluza-Klein idea beyond the five dimensions until the emergence of non-Abelian gauge field theories, invented by Yang and Mills in 1954. In 1963, DeWitt suggested that a unification of Yang-Mills theories and gravitation could be achieved in a higher-dimensional Kaluza-Klein framework. A. Trautman was independently aware of this possibility, as were others. A detailed discussion of the Kaluza-Klein unification of gravity and Yang-Mills theories, including the correct form of the $4 + N$ dimensional metric, first appears in the work of Kerner. The first complete derivation of the four-dimensional gravitational plus Yang-Mills plus scalar theory from a $(4 + N)$-dimensional Einstein-Hilbert action was finally given by Cho and Freund in 1975.

The weakness of this higher-dimensional work was the absence of any good reason why any dimension could compactify, let alone the right number, so as to leave the ordinary four-dimensional "large" world. Although the five-dimensional theory at least admitted the compactified fifth dimension, along with Minkowskian space, as a solution to the five-dimensional equations of motion, even this was not true of the higher-dimensional theories. The essential reason for this difficulty in compacting any dimension is that the higher-dimensional manifolds that give rise to Yang-Mills theories have curvature. If a $4 + N$ dimensional Einstein theory is to compactify into the direct product of four-dimensional space-time M^4 and a compact internal space with isometries, then the metric $\gamma_{mn}(x, y)$ can be written as follows in the zero-mode approximation:

$$
\gamma_{mn}(x, y) =
$$

$$
(15.16) \quad \left[\frac{g_{\mu\nu}(x) + \gamma_{mn}(y)\, \zeta^m_\alpha(y)\, \zeta^n_\beta(y) A^\alpha_\mu(x) A^\beta_\nu(x) \quad \gamma_{mn}(y)\, \zeta^m_\alpha(y) A^\alpha_\mu(x)\gamma_{mn}(x, y)}{\gamma_{mn}(y)\, \zeta^n_\beta(y) A^\beta_\nu(x) \qquad\qquad\qquad \lambda_{mn}(y)} \right].
$$

The metric $\gamma_{mn}(y)$ is that of the corresponding N-dimensional symmetric space, and the upper indices of the Killing vectors $\zeta^n_\alpha(y)$ running over the dimension of the symmetry group. If four-space is to be flat (and, actually, it cannot be flat!), the Ricci tensor $R_{mn} = 0$ for the space-time indices, and therefore $R + \Lambda = 0$. But then R_{mn} must vanish for the internal indices as well, and this cannot be the case if the internal space is curved.

Cremmer and Scherk began to address this problem by pointing out that the inclusion of additional Yang-Mills and scalar matter fields in the higher-dimensional theory would allow classical solutions in which space-time is the direct product of Minkowskian space and a compact internal space of constant curvature. This "spontaneous compactification" was achieved, however, by going beyond the pure Kaluza-Klein framework and including extra fields, in just such a way as to induce the desired compactification. The program of seeking solutions to the combined Einstein-Yang-Mills equations in $4 + D$ dimensions was generalized to a larger class of internal spaces by Luciani, Salam, Duff, and others. All this work on classical, higher-dimensional Kaluza-Klein theories provided a springboard for the study of both Kaluza-Klein supergravity and the quantum dynamics of Kaluza-Klein theories.

Roughly, supergravity is an attempt to unify matter and force as different components of the same agency. This is a kind of supersymmetric theory in which, because the number of Bose and Fermi degrees of freedom have to be equal in supersymmetric theory, Bose fields beyond gravity appear in eleven dimensions. In fact, supersymmetry dictates that the missing Bose degrees of freedom be supplied in the form of a massless antisymmetric tensor field with three indices A_{mnp}, which indeed has $(11 - 2/3) = 84 = 128 - 44$ degrees of freedom. Moreover, in eleven dimensions there exist no matter and no Yang-Mills supermultiplets, so that besides gravity one has only its supersymmetric partner A_{mnp} and gravitino-fields as "matter." The source of gravity is thus fixed by supersymmetry. Furthermore, it is supersymmetry that determines the dimension of spacetime in eleven-dimensional supergravity. Force and matter uniquely determine each other; they are but different components of the same supermultiplet. In ten dimensions, a similar argument can be made, but there we encounter Yang-Mills supermultiplets whose gauge group is fixed, though not uniquely, by the requirement of anomaly cancellation. For superstring theories, similar considerations apply. To find the possible vacuum of the eleven-dimensional theory, we look for a solution of the classical equations in which the eleven-dimensional world manifold M_{11} is of the form $M_{11} = M_d \times M_{11-d}$, where M_d is the space-time and M_{11-d} is the small compact manifold. In the vacuum, we require the spacetime M_d to be maximally symmetric. This then fixes the metric of $M_d(M_{11-d})$. The antisymmetric tensor potential A_{mnp} has its own gauge invariance under

$$(15.17) \quad A_{mnp}(x, y) \rightarrow A_{mnp}(x, y) + \partial_m \alpha_{np}(x, y) + \partial_n \alpha_{pm}(x, y) + \partial_p \alpha_{mn}(x, y)$$

with $\alpha_{mn} = -\alpha_{nm}$. The corresponding gauge invariant quantities are the field strengths F_{mnpr} given by the curl of A_{mnp}, thus

$$(15.18) \quad F_{mnpr} = \partial_m A_{npr} + \partial_n A_{prm} + \partial_p A_{rmn} + \partial_r A_{mnp}.$$

If F or its dual F^* is to have a nonvanishing vacuum expectation value on d-dimensional space-time, without destroying the maximal symmetry, then either d or $11 - d$ must equal the number of indices of F, i.e., $d = 4$ or $d = 7$. In the case $d = 4$, once one has fixed the maximally symmetric form of F, the simplest solutions are obtained by setting $h(y) = 1$ and $F^{mnpq} = 0$. The field equations and Bianchi identities of eleven-dimensional supergravity then require

(15.19) $$F(x, y) = \mathbf{F} = \text{constant,}$$

and the energy momentum tensor of the A-field is equivalent to two cosmological terms, one on M^4 and one on M^7, with cosmological constants of opposite signs. Provided M^4 contains the time dimension, the cosmological constant on M^7 will have the sign appropriate to a compact manifold, so that spontaneous compactification really does occur. M^4 is the maximally symmetric noncompact anti-de Sitter space and M^7 is a compact Einstein space. The scale is set by the expectation value $\mathbf{F}$ of the field strengths.

The rest of four-dimensional physics is determined by the shape of the small seven-dimensional manifold M^7. If, for instance, M^7 is the seven-sphere S^7, then the gauge group in four dimensions is $SO(8)$—the isometry group of S^7—and one finds eight supersymmetries. Just as gauge symmetries in four dimensions are related to the Killing vectors of the small manifold, so the supersymmetries are related to the Killing spinors. All the solutions having one or more surviving supersymmetries are stable with respect to classical perturbations. Some of the solutions lacking any surviving supersymmetry are stable, and others are unstable.

Several problems stand in the way of producing a realistic theory. First of all, the four-dimensional anti-de Sitter space has too large a cosmological constant, which has somehow to be eliminated. Of course, Higgs mechanisms in four dimensions further affect the cosmological constant, and it is the end product that must be very small or zero. Another serious problem is the lack of chiral fermions, at least so long as bound states and solitons are ignored. It seems that these problems can be solved in the context of higher-dimensional superstring theory, which recently demonstrated the possibility of a finite quantum theory of gravity, whereas the eleven-dimensional supergravity theory, although trivially finite at one loop, is questionable in this regard.

Chapter 16

Creation of Universes
from Nothing

The standard hot cosmological model of the universe gives a successful description of many features of its evolution. But the model is not totally satisfactory, for it requires that the Big Bang be preceded by rather unnatural initial conditions. One has to postulate that the universe began in a homogeneous and isotropic state with tiny density fluctuations that would evolve into galaxies. Homogeneity and isotropy must extend to scales far exceeding the causal horizon at the Planck time. Moreover, the energy density of the universe must be tuned to near its critical level, with an accuracy of an incredible $\sim 10^{-55}$.

In the last few years, there has been a growing hope that these initial conditions can be explained as resulting from physical processes in the very early universe. Guth (1981) has suggested that the homogeneity, isotropy, and flatness puzzles can be solved if the universe had passed through a de Sitter phase of exponential expansion (inflation) in its early history [$a(t) \propto \exp(Ht)$, where $a(t)$ is the scale factor]. Such a phase can arise in a first-order phase transition with strong supercooling. It has been suggested [Witten (1981); Albrecht and Steinhardt (1982)] that extreme supercooling can occur in grand unified models with the Coleman-Weinberg type of symmetry breaking. Initially, it was not clear how to end the exponential expansion and return to a radiation-dominated universe.

A plausible answer has emerged recently. At some temperature T_0, the false vacuum becomes unstable, owing to thermal or gravitational effects. The Higgs field ϕ starts rolling down the effective potential toward the absolute minimum, $\phi = \sigma$. The Coleman-Weinberg potential is very flat for small values of ϕ ($\phi \ll \sigma$), and the typical rollover time, τ, can be much greater than the expansion time, H^{-1}. Until ϕ becomes of the order of σ, exponential expansion continues, and the scale of the universe grows by a factor of $\sim \exp(H\tau) \ggg 1$. To solve the homogeneity and flatness problems, we need $\exp(H\tau) \geq 10^{28}$. Most of this growth takes place after the destabilization of the false vacuum. When ϕ becomes $\sim \sigma$, the vacuum energy thermalizes, and the universe enters a radiation-dominated period. The baryon number can be generated during the thermalization or shortly afterward. Density fluctuations can be generated by vacuum strings produced at a later phase transition. Another attractive feature of this scenario is that the problem of superabundance of heavy magnetic monopoles does not arise: the Higgs expectation value is uniform over the whole visible universe.

Now that we have a plausible ending to the inflationary scenario, we can start wondering about its beginning, where the situation remains rather depressing. There is a cosmological singularity at $t = 0$, and the origin of the initial thermal state is mysterious. Besides, there is another problem if we assume that the universe is closed (which seems to be a more aesthetically appealing choice). It is natural to assume that at Planck time ($t \sim t_p$) the size and energy density of the universe are $O(1)$ in Planck units. But then the universe will expand and recollapse in about one Planck time, its size will never much exceed the Planck length, and the phase of exponential expansion will never be reached (assuming that the grand unification mass scale is much smaller than the Planck mass, $\sigma \ll m_p$). In order to cool down to temperatures $\sim 10^{14}$ GeV, the energy density at $t \sim t_p$ must be tuned to be near the critical density with an accuracy of $\sim 10^{-10}$. This is just a milder version of the same flatness problem that we have faced before.

In a new cosmological scenario, suggested in the 1980s by A. Vilenkin, A. H. Guth, and P. J. Steinhardt, the universe is spontaneously created from literally *nothing*, and is therefore free from the difficulties mentioned in the preceding paragraph. This scenario requires no changes in the fundamental equations of physics; it offers only a new interpretation to a well-known cosmological solution. It must be recalled, however, that E. P. Tryon (1973) had already proposed a big bang model in which our universe is a fluctuation of the vacuum, in the sense of quantum field theory. His model predicts a universe that is homogeneous, isotropic, and closed, and consists equally of matter and antimatter. All these predictions seem to be supported by, or consistent with, present observations.

Another important work that should be mentioned here is that of E. Streeruwitz (1975). He calculated the vacuum expectation value of the energy-momentum tensor of a quantized scalar field in a Robertson-Walker universe. The resulting divergences are regularized by averaging over an appropriate mass spectrum, which fulfills the same regularization conditions as those needed for regularization of the vacuum energy-momentum tensor in Minkowskian space. Up to higher orders in time derivatives and an inverse radius of the universe, there are three contributions to the vacuum energy-momentum density: a cosmological term, a term proportional to the Einstein tensor, and a term derivable by variation of the gravitational Lagrangian containing quadratic expressions of the curvature quantities R, $R_{\mu\nu}$, $R_{\mu\nu\rho\sigma}$. These contributions are estimated with the following result: there is no direct evidence that the collapse of a Robertson-Walker universe is averted at some realistic dimensions of the universe, or that gravitation is described in a natural way by the elastic properties of the vacuum in which some particles interact. But there is some evidence that certain types of quantum corrections must be taken into account in general relativity and are important at least for highly collapsed states of the early universe.

Following previous works of Guth, Steinhardt, and Linde, Vilenkin considered a model of interacting gravitational and matter fields. The matter content of the model can be taken to be that of some grand unified theory (GUT). The absolute minimum of the effective potential is reached when the Higgs field ϕ responsible for the symmetry breaking of the GUT acquires a vacuum expectation value, $\langle\phi\rangle = \sigma \ll m_p$. The symmetrical vacuum state, $\langle\phi\rangle = 0$, has a nonzero energy density, ρ_v. For a Coleman-Weinberg potential,

(16.1)
$$\rho_v \sim g^4 \, \sigma^4,$$

where g is the gauge coupling.

Suppose now that the universe begins in the symmetrical vacuum state and is described by a closed Robertson-Walker metric

(16.2)
$$ds^2 = dt^2 - a^2(t) \, [dr^2/(1 - r^2) + r^2 d\Omega^2].$$

The scale factor $a(t)$ can be found from the evolution equation

(16.3)
$$\dot{a}^2 + 1 = (8\pi G/3)\rho a^2,$$

where $\dot{a}^2 = da/dt$. Note that the Euclidean version of equation (16.3) is $-a^2 + 1 = H^2 a^2$, and the solution is

(16.4)
$$a(t) = H^{-1} \cos(Ht).$$

The solution to equation (16.3) is the de Sitter space,

(16.5)
$$a(t) = H^{-1} \cosh(Ht),$$

where

$$H = \left(\frac{8\pi G \rho_v}{3} \right)^{\frac{1}{2}}.$$

It describes a universe that is contracting at $t < 0$, reaches its minimum size $(a_{min} = H^{-1})$ at $t = 0$, and is expanding at $t > 0$. This behavior is analogous to that of a particle bouncing off a potential barrier at $a = H^{-1}$. (Here a plays the role of the particle coordinate.) We have seen in the preceding chapter that particles can tunnel through potential barriers, which suggests that the birth of the universe might be a quantum tunneling effect. Then the universe will have emerged having a finite size $(a = H^{-1})$ and zero "velocity" $(a = 0)$; its following evolution is described by equation (13.5) with $t > 0$.

Equations (16.2) and (16.4) describe a four-sphere S^4. This is the well-known de Sitter instanton. The solution of (16.4) does bounce at the classical turning point $(a = H^{-1})$, but it does not approach any initial state at $t \to \pm\infty$. In fact, S^4 is a compact space, and the solution (16.4) is defined only for $|t| < \pi/2\,H$. The instanton (16.4) can be interpreted as describing the tunneling to de Sitter space (16.5) from *nothing*. The birth of the universe is symbolically represented in fig. 25, de Sitter space-time in fig. 26.

An *instanton* is a classical solution to equations of motion with a finite, non-zero action, either in quantum mechanics or in quantum field theory. More precisely, it is a solution to the equations of motion of the classical field theory on a Euclidean space-time. In such a theory, solutions to the equations of motion may be thought of as critical points of the action. These critical points may be local maxima of the action, local minima, or saddle points. An instanton can be used to calculate a certain transition probability that a quantum-mechanical particle will tunnel through a region of potential energy. It may be that the easiest example for a system with an instanton effect is the particle in a double-well potential. In contrast to a particle in classical mechanics, there is a nonvanishing probability for it to cross a region of potential energy higher than its own energy. In the path integrals approach to instantons, the potential gets rotated by 180 degrees, thus standing on its head, exhibiting two "hills" of maximal energy.

Results obtained from the mathematically well-defined Euclidean path integral may be Wick-rotated back and give the same physical results as would be obtained by appropriate treatment of the (potentially divergent) Minkowskian path integral. In fact, calculating the transition probability that the particle will tunnel through a classically forbidden region $(V(x))$ with the Minkowskian path integral corresponds to calculating the transition probability for tunneling

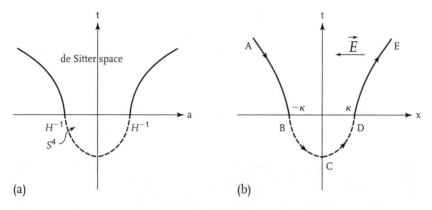

Fig. 25. A schematic representation of (a) the birth of the inflationary universe and (b) the creation of electron–positron pairs in a constant electric field E. In both cases, the dashed semicircles represent the "under-barrier" part of the circular trajectory. (Below the horizontal axis t is the Euclidean time.) The classical evolution starts at $t = 0$.

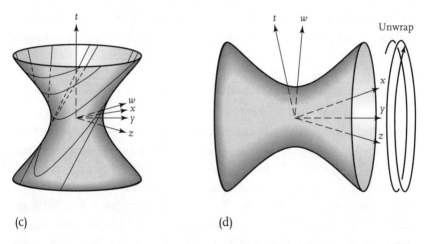

Fig. 26. In (c), the de Sitter space-time is pictured as a hyperboloid, with two spatial dimensions suppressed. It is a Lorentzian "4-sphere" (of imaginary radius, giving the intrinsic metric signature $+---$) in Minkowskian 5-space $\mathbb{M}^5$ (whose metric is $ds^2 = dt^2 - dw^2 - dx^2 - dy^2 - dz^2$). To arrive at the steady-state model, we "cut" the hyperboloid in half, along $t = w$; constant time is given by constant positive $t - w$. In (d), anti-de Sitter space-time (pictured as a hyperboloid with two spatial dimensions suppressed) is a Lorentzian "4-sphere" (of positive radius, giving an intrinsic metric signature $+---$), in pseudo-Minkowskian 5-space (with metric $ds^2 = dt^2 + dw^2 - dx^2 - dy^2 - dz^2$). As defined, we have closed timelike curves, but these can be removed by infinitely "unwrapping" them in the (t, w)-plane.

through a classically allowed region (with potential $-V(x)$) in the Euclidean path integral. (Pictorially speaking—in the Euclidean picture—this transition corresponds to a particle rolling from one hill of a double-well potential, then standing on its head on the other hill.) This classical solution of the Euclidean equations of motion, often named "kink solution," is an example of an instanton. In this example, the two "vacua" of the double-well potential turn into hills in the Euclideanized version of the problem. Thus, the instanton field solution of the $(1 + 1)$-dimensional field theory, which was the first quantized quantum-mechanical system, could be interpreted as a tunneling effect between the two vacua of the physical Minkowskian system.

Instantons are important in quantum field theory (QFT) because (a) they appear in the path integral as the leading quantum corrections to the classical behavior of a system, and (b) they can be used to study the tunneling behavior in various systems, such as that of a Yang-Mills theory. Instantons illustrate the vacuum structure of QFT: the true vacuum of a field theory may be an "overlap" of several topologically inequivalent sectors, so called "topological vacua." For a Yang-Mills theory these inequivalent sectors can be (in an appropriate gauge) classified by the third homotopy group of $SU(2)$ (whose group manifold is the 3-sphere S^3). A certain topological vacuum (a "sector" of the true vacuum) is labeled by a topological invariant, the Pontryagin index. As the third homotopy group of S^3 has been found to be the set of integers $\pi_3(S^3) = \mathbb{Z}$, there are infinitely many topologically inequivalent vacua, denoted by $|N\rangle$, where N is their corresponding Pontryagin index.

An instanton is a field configuration fulfilling the classical equations of motion in Euclidean space-time, which is interpreted as a tunneling effect between these different topological vacua. It is again labeled by a whole number, its Pontryagin index, Q. One could use an instanton with index Q to quantify tunneling between topological vacua $|N\rangle$ and $|N + Q\rangle$. If $Q = 1$, the configuration is named BPST instanton after its discoverers, A. A. Belavin, A. Polyakov, A. S. Schwartz, and Yu. S. Tyupkin. The true vacuum of the theory, labeled by an "angle" θ, is an overlap of the topological sectors: $|\theta\rangle = \sum_{N=-\infty}^{N=+\infty} e^{i\theta N}|N\rangle$. The Yang-Mills gauge-field configuration of an instanton is quite different from that of the vacuum. Consequently, instantons cannot be studied by using Feynman diagrams, which include only perturbative effects. Instantons are fundamentally nonperturbative.

The concept of the universe being created from nothing is an astonishing one. Fig. 26, above, illustrates how the instanton solution describing the creation of a pair is obtained from the equation

(16.6) $$x - x_0 = \pm \, [\kappa^2 + (t - t_0)^2]^{1/2},$$

(where $\kappa = |m/eE|$ and $x_0, t_0 = $ const.; the classical turning points are at $x = x_0 \pm \kappa$) by changing t to $-it$:

(16.7) $$(x - x_0)^2 + (t - t_0)^2 = \kappa^2.$$

which describes a circular trajectory; that is, again, we have a compact instanton. The process of pair creation is symbolically represented above, in fig. 25(b). AB and DE are classically allowed trajectories. AB describes an electron moving backwards in time, that is, a positron. The semicircle BCD represents the instanton. The instanton solution (equation 16.8 below) can be used to estimate the semiclassical probability P of pair creation per unit length per unit time: $P \propto \exp(-S_E)$, where S_E is the Euclidean action,

(16.8) $$S_E = \int [m(1 + x^2)^{1/2} - eEx] \, dt.$$

Obviously, the evaluation of the probability P is made possible by the pair creation that takes place in a background flat space. The instanton solution contributes to the imaginary part of the vacuum energy. But such a calculation does not make sense for our de Sitter instanton: it is silly to evaluate the imaginary part of the energy of nothing. The only relevant question seems to be whether or not the spontaneous creation of universes is possible. The existence of the instanton (equation 16.4) suggests that it is. One can assume, as usual, that instantons, being stationary points of the Euclidean action, make a dominant contribution to the path integral of the theory.

There may be several relevant instanton solutions. For example, we can have a de Sitter instanton with broken grand unified symmetry, but unbroken Weinberg-Salam symmetry. Then the vacuum energy is $\sim\rho_v \sim \sigma_{WS}^4 \ll \rho_v$, where $\sigma_{WS} \sim 100\,\mathrm{GeV}$ is the energy scale of the $SU(2) \times U(1)$ symmetry breaking. The Euclidean action of a de Sitter instanton is negative, $S_E = -3m_p^4/8\rho_v$. If one assumes that the instanton with the smallest value of S_E corresponds, in some sense, to most probable universes, then most of the universes never heat up to temperatures greater than $100\,\mathrm{GeV}$ and have practically vanishing baryon numbers. Obviously, we must be living in one of the rare universes that tunneled to the symmetrical vacuum state.

It remains to talk about what happens to the universe after the tunneling. The symmetrical vacuum state is not absolutely stable. It can decay by quantum tunneling, or it can be destabilized by quantum fluctuations of the Higgs field. The Higgs field starts rolling down the effective potential toward the celebrated ending of the inflationary scenario. When the vacuum energy thermalizes, the universe heats up to a temperature $T^* \sim \rho_v^{1/4}$. In Vilenkin's model, this is the maximum temperature the universe has ever had. In principle, the only verifiable prediction of the model is that the universe must be closed. Guth, however, has argued that the inflationary scenario almost certainly overshoots, so that $\rho = \rho_{crit}$ with a very high accuracy even at the present time. This means that we shall have to wait a long time until the sign of $(\rho = \rho_{crit})$ can be determined experimentally. The advantages of the scenario proposed by A. Vilenkin (1982, 1984, 1998) are also of an aesthetic nature. The scenario offers a cosmological model that does not have a singularity at the big bang (there may still be a final singularity) and does not require any initial or boundary conditions. The structure and evolution of the universe(s) are completely determined by the laws of physics.

It is worth noting here the profound relationship that seems to exist between the quantum theory of black holes and the creation of particles pairs[1] [on this subject, see Hawking (1975), and Hawking and Turok (1998)]. As we have seen before, one can create pairs of positively and negatively charged particles in a strong electric field. One way of looking at this is to note that in flat Euclidean space a particle of charge q, such as an electron, would move on a circle in a uniform electric field E (fig. 27). One can analytically continue this motion from the imaginary time τ to real time t. One derives pairs of positively and negatively charged particles accelerating away from each other, then pulled apart by the electric field (fig. 28). The process of pair creation is described by chopping the two diagrams in half along the $t = 0$ or $\tau = 0$ lines. One then joins the upper half of the Minkowskian space diagram to the lower half of the Euclidean space diagram (fig. 29). This presents a picture in which the positively and negatively charged particles are really the same particle. It tunnels through Euclidean space to get from one Minkowskian space worldline to the other. To a first approximation, the probability for pair creation is e^{-1}, where the Euclidean action is given by

(16.9) $S_E = 2\pi m^2 / qE.$

Pair creation by strong electric fields has been observed experimentally, and the rate agrees with these estimates.

Following Hawking's reasoning, black holes can also carry electric charges, so one might expect that they could also be pair created. But the rate would be tiny compared to that for electron-positron pairs because the mass-to-charge ratio is 10^{20} times bigger. This means that any electric field would be neutralized

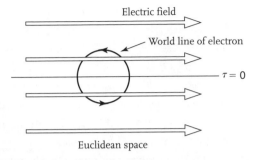

Fig. 27. In Euclidean space, an electron moves on a circle in an electric field.

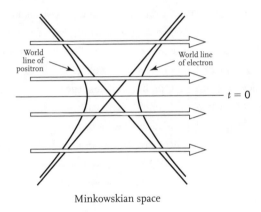

Fig. 28. In Minkowskian space, we obtain a pair of oppositely charged particles accelerating away from each other.

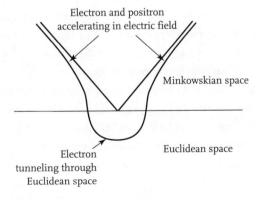

Fig. 29. Pair creation is described by joining half of the Euclidean diagram to half of the Minkowski diagram.

by electron-positron pair creation long before there was a significant probability of a pair's creating black holes. Neither could such black holes be produced by gravitational collapse, because there are no magnetically charged elementary particles. But one might expect that they could be pair created in a strong magnetic field. In this case, there would be no competition from ordinary particle creation, because ordinary particles do not carry magnetic charges. So the magnetic field could thus become strong enough to yield a significant change in the creation of a pair of magnetically charged black holes.

A solution found in 1976 represents two magnetically charged black holes accelerating away from each other in a magnetic field (fig. 30). If one analytically continues the acceleration to imaginary time, one has a picture very like that of the electron pair creation (fig. 31). The black hole moves on a circle in a curved Euclidean space, just as the electron moves in a circle in flat Euclidean space.

There is a complication in the black hole case. The imaginary-time coordinate is periodic about the black hole's horizon, as well as about the center of the circle on which the black hole moves. One has to adjust the mass-to-charge ratio of the black hole to render these periods equal. Physically, this means that one chooses the parameters of the black hole such that its temperature is equal to the temperature it sees, because it is accelerating. The temperature of a magnetically charged black hole tends to zero as the charge tends to the mass in Planck units. Thus, for weak magnetic fields, and hence low acceleration, one can always match the periods.

As in the case of pair creation of electrons, one can describe pair creation of black holes by joining the lower half of the imaginary-time Euclidean solution to the upper half of the real-time Lorentzian solution (fig. 32). One can think of the black hole as tunneling through the Euclidean region and emerging as a pair of oppositely charged black holes that accelerate away from each other, pulled apart by the magnetic field. The accelerating black hole solution is not asymptotically flat, because it tends to a uniform magnetic field at infinity. But one can nevertheless use it to estimate the rate of pair creation of black holes in a local region of magnetic field. One could imagine that after being created, the black holes move far apart into regions lacking magnetic field. One could then treat each black hole separately, as a black hole in asymptotically flat space. One could also throw an arbitrarily large amount of matter and information into each hole. The holes would then radiate and lose mass, but they could not lose magnetic charge because there are no magnetically charged particles. Thus, they would eventually get back to their original state, with the mass slightly greater than the charge.

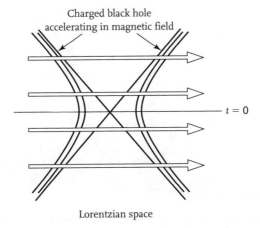

Fig. 30. A pair of oppositely charged black holes accelerating away from each other in a magnetic field.

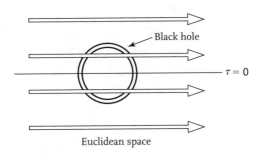

Fig. 31. A charged black hole moving on a circle in Euclidean space.

One could then bring the two holes back together again and let them annihilate each other.

The annihilation process can be regarded as the time reverse of the pair creation. Thus it is represented by the top half of the Euclidean solution joined to the bottom half of the Lorentzian solution. In between the pair creation and the annihilation, one can have a long Lorentzian period in which the black holes move far apart, accrete matter, radiate, and then come back together again. But the topology of the gravitational field will be the topology of the Euclidean-Ernst solution. This is $S^2 \times S^2$ minus a point (fig. 33). See also figs. 34 and 35.

It is believed by Hawking and other physicists that the black hole thermodynamics cannot just be a low energy approximation. They believe that gravitational

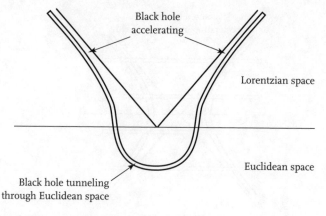

Black hole
accelerating

Lorentzian space

Euclidean space

Black hole tunneling
through Euclidean space

Fig. 32. Tunneling to produce a pair of black holes is also described by joining the
Euclidean diagram to half of the Lorentzian diagram.

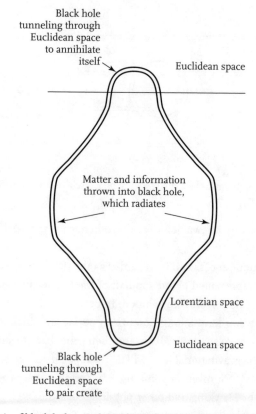

Black hole
tunneling through
Euclidean space
to annihilate
itself

Euclidean space

Matter and information
thrown into black hole,
which radiates

Lorentzian space

Euclidean space

Black hole
tunneling through
Euclidean space
to pair create

Fig. 33. A pair of black holes produced by tunneling, and eventually annihilated
again by tunneling.

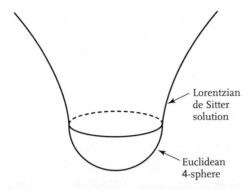

Fig. 34. The tunneling to produce an expanding universe is described by joining half of the Euclidean solution to half of the Lorentzian solution.

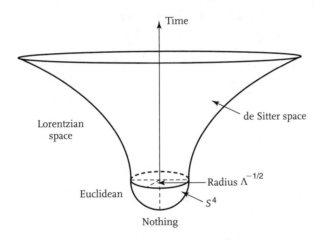

Fig. 35. The creation, from nothing, of a closed de Sitter universe.

entropy will not disappear even if we must proceed to a more fundamental theory of quantum gravity. One can see from the previous gedankenexperiment (see chapter 4) that one acquires intrinsic gravitational entropy and loss of information when the topology of space-time differs from that of flat Minkowskian space. If the black holes that are pair created are large relative to the Planck size, the curvature outside the horizons will be everywhere small compared to the Planck scale. If information is lost in macroscopic black holes, it should also be lost in processes in which microscopic, virtual black holes appear because of quantum fluctuations of the metric. One could imagine that particles and information could fall into these holes and get lost. Quantities like energy and electric charge that are coupled to gauge fields would be conserved, but other information and global charge would be lost.

This outcome would have far-reaching implications for quantum theory. It is normally assumed that a system in a pure quantum state evolves in a unitary way through a succession of pure quantum states. But if there is loss of information through the appearance and disappearance of black holes, there cannot be a unitary evolution. Instead, the loss of information will mean that the final state after the black holes have disappeared will be what is called a *mixed quantum state*. This can be regarded as an ensemble of different pure quantum states, each with its own probability. But because the mixed quantum state is not with certainty in any one state, one cannot reduce the probability of the final state to zero by interfering with any quantum state. This means that gravity introduces a new level of unpredictability into physics, over and above the uncertainty usually associated with quantum theory.

Chapter 17

String Landscape and Vacuum Energy

*The Emergence of a Multidimensional World
from Geometrical Possibilities*

Let us examine in more detail the concluding idea of chapter 16. One possible starting point of the string picture is that the solution to Einstein's field equations (roughly speaking, they say that matter tells space-time how to curve, and space-time tells matter how to move) is not unique, and many different geometries are therefore allowed. The case of 5-dimensional Kaluza-Klein geometry provides a simple example of this non-uniqueness (the central theme of Kaluza-Klein theory is that the physical laws we see depend mainly on the geometry of hidden extra dimensions). The circumference of the small dimension can take any size at all: in the absence of matter, four large flat dimensions, plus a circle of any size, solve Einstein's equations. (Similar multiple solutions also exist when matter is present.)

In string theory, we have several extra dimensions, which results in many more adjustable parameters. One extra dimension can be wrapped up only in a circle. When more than one extra dimension exists, the bundle of extra dimensions can have many different shapes—i.e., topologies—such as a sphere, a doughnut, two doughnuts joined together, and so on. Each doughnut loop (a "handle") has a length and a circumference, resulting in a huge assortment of possible geometries for the small dimensions. In addition to the handles, there are parameters that correspond to the locations of branes[1] and the different amounts of flux wound around each loop. Any given solution to the equations of string theory represents a specific configuration of space and time. In particular, it specifies the arrangement of the small dimensions, along with their associated brane and the lines of force known as flux lines (fig. 36). Our world has six extra dimensions, so every point of our familiar 3-dimensional space hides an associated tiny 6-dimensional space, or manifold—a 6-dimensional analogue of the circle.

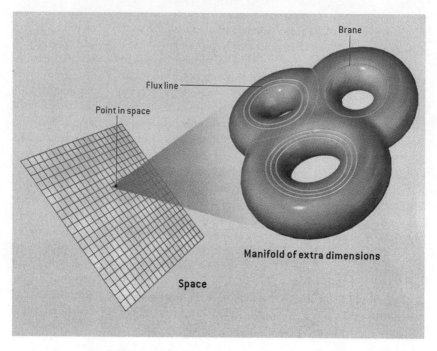

Fig. 36. The hidden space. By permission of Don Foley.

The physics that is observed in the three large dimensions depends on the size and the structure of the manifold: how many doughnut-like "handles" it has, the length and circumference of each handle, the number and locations of its branes, and the number of flux lines wrapped around each doughnut.

Yet the solutions composing this vast collection are not all equal: each configuration has a potential energy, contributed by fluxes, branes, and the curvature itself of the curled-up dimensions. This energy is called the *vacuum energy*, because it is the energy of the space-time when the four large dimensions are completely devoid of matter or fields. The geometry of the small dimensions will try to adjust to minimize this energy, just as a ball placed on a slope will start to roll downhill to a lower position. To understand what consequences follow from this minimization, we must focus first on a single parameter: the overall size of the hidden space. We can plot a curve showing how the vacuum energy changes as this parameter varies. An example is shown in the illustration below (fig. 37). At very small sizes, the energy is high, and the curve starts out high at the left. Then, from left to right, it dips down into three valleys, each one lower than the preceding one. Finally, at the right, after climbing out of the last valley, the curve trails off down a shallow slope to a constant value. The bottom of the leftmost valley is above zero energy; the middle one is at exactly zero; and the righthand one is below zero. These three local minima differ by virtue of whether the resulting vacuum energy is positive, negative, or zero. In our universe, the size of the hidden dimensions is not changing with time: if it were, we would see the constants of nature changing. Thus, we must be sitting at a minimum. In particular, we seem to be sitting at a minimum with a slightly positive vacuum energy.

Because there is more than one parameter, we should actually think of this vacuum-energy curve as one slice through a complex, multidimensional mountain range: the landscape of string theory (fig. 38). The minima of this multidimensional landscape—the bottoms of depressions where a ball comes to rest—correspond to the stable configurations of space-time (including branes and fluxes), which are called *stable vacua*. The landscape of string theory is very complicated (much more than the usual one counting only two independent directions: north-south and east-west) with hundreds of independent directions. The landscape dimensions should not be confused with the actual spatial dimensions of the world; each axis measures not some position in physical space but some aspect of the geometry, such as the size of a handle or the position of a brane. The landscape of string theory is far from being fully mapped out. Calculating the energy of a vacuum state is a difficult problem, and usually depends on finding suitable approximations. We cannot be sure how many stable vacua there are—that is, how many points where a ball could rest. But the number could very well be enormous. Some research suggests that there are solutions with up to about 500 handles, but not many more. We can wrap different numbers of flux lines around each handle, but not too many, because they would make the space unstable, like the righthand part of the curve in fig. 39. If we

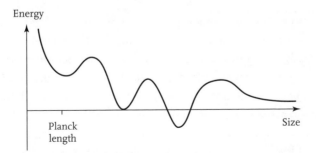

Fig. 37. A string-theory landscape and a topography of energy.

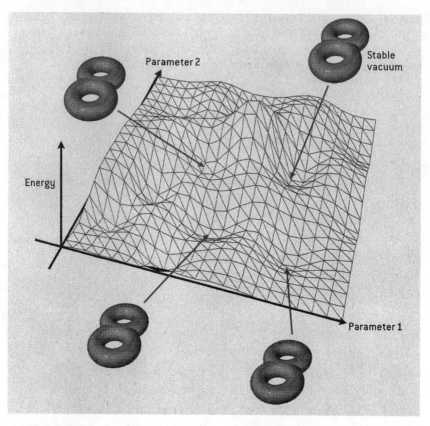

Fig. 38. A topographical illustration of a string-theory landscape having a vast number of dimensions. We represent it by a landscape showing the variation of the energy contained in empty space when only two features change. The manifold of extra dimensions tends to terminate at the bottom of a valley, which is a stable string solution, or a stable vacuum. The blue regions are below zero energy.

By permission of Don Foley.

suppose that each handle can have from zero to nine flux lines (ten possible values), then there would be 10^{500} possible configurations. Even if each handle could have only zero or one flux unit, there are 2^{500}, or about 10^{150}, possibilities. As well as affecting the vacuum energy, each of the many solutions will conjure up different phenomena in the 4-dimensional macroscopic world, by defining which kinds of particles and forces are present, and what masses and interaction strengths they have. String theory may come to provide us with a unique set of fundamental laws, but the laws of physics that we perceive in the macroscopic world will depend on the geometry of the extra dimensions.

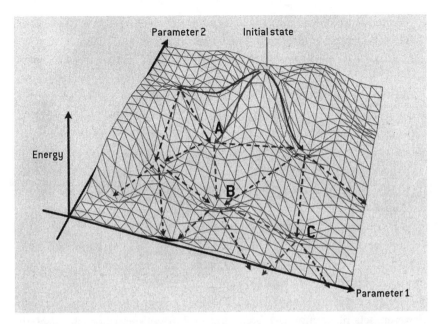

Fig. 39. Quantum space configurations. Quantum effects allow a manifold to change state abruptly at some point—to tunnel through the intervening ridge to a nearby lower valley. We see how one region of the universe (*solid arrows*) might evolve, by starting out at a high mountaintop, rolling down into a nearby valley (*vacuum A*), eventually tunneling through to another, lower valley (*vacuum B*), and so on. Different regions of the universe will randomly follow different paths. The effect suggests an infinite number of explorers traversing the landscape, passing through all possible valleys (*dashed arrows*). By permission of Don Foley.

But many profound questions about the string-theory landscape remain to be answered. Which stable vacuum describes the physical world we experience? Why has nature adopted this particular vacuum and not any other? Have all other solutions been demoted to mere mathematical possibilities, never to become true? Instead of reducing the landscape to a single chosen vacuum, some physicists have recently proposed a very different picture, based on two important ideas. The first is that the world need not be stuck with one lone configuration on the small dimensions forever, because a rare quantum process allows the small dimensions to jump from one configuration to another. The second idea is that Einstein's general relativity theory, which is a component of string theory, implies that the universe can grow so rapidly that different configurations will coexist side by side in different sub-universes, each large enough to be unaware of the others.

As mentioned before, each stable vacuum is characterized by its numbers of handles, branes, and flux quanta. But one can also take into account the fact that each of these elements can be created and destroyed, so that after periods of stability, the world can snap into a different configuration. In our theoretical landscape picture, the disappearance of a flux line or other change of topology is a quantum jump over a mountain ridge into a lower valley. Consequently, as time goes on, different vacua might come into existence. Suppose that each of the 500 handles in our earlier example starts out with nine units of flux. One by one, the 4,500 flux units will decay, in some sequence governed by the probabilistic predictions of quantum theory, until all the energy stored in the fluxes is used up. We start in a high mountain valley and leap randomly over the adjoining ridges, visiting 4,500 successively lower valleys. We are led through some varied scenery, but we pass by only a minuscule fraction of the 10^{500} possible solutions. A key part of the picture is the effect of the vacuum energy on how the universe evolves. Ordinary objects such as stars and galaxies tend to slow down an expanding universe, and can even cause it to recollapse. Positive vacuum energy, however, acts like antigravity: according to Einstein's equation, it causes the three dimensions that we perceive to grow more and more rapidly. This rapid expansion has an important and surprising effect when the hidden dimensions tunnel to a new configuration.

Remember that at every point in our 3-dimensional space there sits a small 6-dimensional space, which lives at some point on the landscape. When this small space jumps to a new configuration, the jump does not happen at the same instant everywhere. The tunneling first happens at one place in the 3-dimensional universe, and then a bubble of the new low-energy configuration expands rapidly. If the three large dimensions were not expanding, this growing bubble would eventually overrun every point in the universe. But the old region is also expanding, and this expansion can easily be faster than that of the new bubble. Because the original configuration keeps growing, eventually it will decay again at another location, to another nearby minimum in the landscape. The process will continue infinitely many times, decays happening in all possible ways, with far separated regions losing fluxes from different handles. In this manner, every bubble will be host to many new solutions. Instead of a single sequence of flux decay, the universe thus experiences all possible sequences, resulting in a hierarchy of nested bubbles, or sub-universes. The result is quite similar to the eternal inflation scenario proposed by A. Guth, A. Vilenkin, and A. Linde.

The picture we have described explains how all the different stable vacua of the string landscape come into existence at various locations in the universe, thus forming innumerable sub-universes. This result may solve one of the most important and long-standing problems in theoretical physics—one related to the vacuum energy. To Einstein, what we now think of as vacuum energy was an arbitrary mathematical term—a "cosmological constant"[2]—that could be added to his equation of general relativity to make it consistent with his conviction that the universe is static. To obtain a static universe, he proposed that this constant takes a positive value, but he abandoned the idea after observations proved the universe to be expanding. With the advent of quantum field theory, empty space—the vacuum—became a busy place, full of virtual particles and fields popping in and out of existence, each particle and field carrying some positive or negative energy. Concerning the nonzero value of the vacuum energy, the general idea of string theory is that the complicated geometries of hidden dimensions might produce a spectrum for vacuum energy that includes values in the experimental window. If the landscape picture is valid, a nonzero vacuum energy should be observed, most likely not much smaller than $10^{-118}\Lambda p$. (The value of Λp is about 10^{94} grams per cubic centimeter, or one Planck mass per cubic Planck length.)

Chapter 18

Concluding Remarks

In these chapters, I have tried to stress the point that relativistic quantum field theory makes a very clear distinction between what we would intuitively understand to be an *absolute* void and what we experience as the vacuum of space. And as we had occasion to remark, even in the deepest reaches of space we will find an atom or molecule here or there, and photons of energy flying through continuously, at the speed of light. But there is also a quantum potential that exists at every point in the vacuum of our 3-dimensional physical space. Under the proper conditions, matter and energy can literally be made to materialize out of what we used to think of as nothing.

Of course, this does not mean that any conceivable fantasy can be "popped" out of space with ease. The conditions necessary to create anything more than a scattering of subatomic particles would require extremely high energies and a control of the process far beyond any practical ability, at least within our present means. Nevertheless, according to the present mathematical models of physical theories, it is *scientifically possible* to create anything right out of the vacuum. The classical notion of empty space has obviously been abandoned. According to the physicist John A. Wheeler (1968), "No point is more central than this, that empty space is not empty. It is the seat of the most rich and surprising physics."

The quantum vacuum is thought of as a seething froth of real particle-virtual particle pairs going in and out of existence, both continuously and rapidly. As we have seen, each of these strange pairs consists of a particle and its antiparticle, one of which has a *negative* energy and is thus called "virtual." Out of a singularity in space, which by definition really is nothing, each pair simply comes into existence. Why? Because the probability exists, because the universe is open to many (perhaps infinite) possibilities that had not previously been predicted, but nonetheless could occur in the future. The future, in other words, seems to be intrinsically unpredictable, at least according to quantum mechanics but also to chaos theory[1] and other system-dynamics theories. It is impossible to predict in any way, from any a priori information, which direction or trajectory a certain physical entity or event will follow (proceed along), or which it will not follow. These considerations imply that physics has, in a way, given up, if its original purpose was to know enough that, given our present circumstance we can predict (with precision and certainty) what will happen next.

This (likely) intrinsically indeterminate and in some sense fuzzy character of the quantum laws of nature (at the Planck scale) represents a strong denial of the philosophical idea underlying the deterministic vision of the natural phenomena, according to which it is necessary for the very existence of science that the same conditions always produce the same results. And it is not merely our ignorance of the internal mechanisms, and of the internal complications, that makes nature appear to embrace uncertainty and a horde of possibilities. The uncertainty seems somehow to be intrinsic. Nature herself does not know which way its constituents are going to go.

The fact that this view of reality arises simply from the possibility that it *can*, sounds both contrived and fantastic, but it is nevertheless a very real reality. It is rooted, as we have said, in the uncertainty principle. The quantum potential of the vacuum, revealed dramatically by the giant particle accelerators in the emergence of new and exotic particles, is an excellent example of the fact that the strange implications of quantum theory constitute a great deal more than just a temporary limitation of our contemporary mathematical schemes. The "strangeness" reflects an actual view of a reality that does not (necessarily and always) conform to everyday commonsense logic. Virtual particles with negative energy, just like antiparticles with opposite charge, can actually exist, at least for a tiny fraction of a second, as if they were perfect "mirror images" of their normal partners. Inside a black hole, they can even become *real* particles.

The notion of quantum vacuum gives you an idea of what today's quantum field theory is like, and it is a good illustration of what modern physics considers quanta to be. Quanta are not "things in space" so much as they are discrete manifestations of the underlying fields that pervade all of space, even where no quanta appear to be. From here, it is not such a leap to consider one of the strangest revelations of quantum field theory: that quanta actually appear to be completely *independent of space-time separation*. For example, a quantum can disappear at one point in space-time only to instantly reappear somewhere else entirely (faster than the speed of light, if thought to have actually gone through space-time).

This exceedingly unconventional and logic-boggling behavior, called the tunnel effect, is very real.

Let us, then, sum up the most important points we have addressed here:

1. In quantum field theory (QFT), a *quantum fluctuation* is a momentary change in the amount of energy at a point in space. The creation of new forms of energy, as we saw in previous chapters, arises from the intrinsic content of Heisenberg's uncertainty principle. According to one formulation of the principle, energy and time can be related by the relation $\Delta E \Delta t \approx h/2\pi$. This implies that conservation of energy can appear to be violated, but only for fleeting moments. This allows the creation of particle-antiparticle pairs of virtual particles. The effects of these particles are measurable, for example, in the effective charge of the electron, different from its "naked" charge. Quantum fluctuations may have been important in the origin of the structure of the universe: according to the model of inflation, the fluctuations that existed when inflation began were amplified, forming across time the seed of all current observed structure.

2. In QFT, and specifically quantum electrodynamics (QED), *vacuum polarization* describes a process by which background electromagnetic field produces virtual electron-positron pairs that alter the distribution of the charges and currents that generated the original electromagnetic field. Vacuum polarization is also sometimes referred to as the self-energy of the gauge boson (photon). The true vacuum, i.e., the ground state of the interacting theory (in which electrons, positrons, and photons do interact), contains short-lived "virtual" particle-antiparticle pairs that are created out of the Fock vacuum and then annihilate each other. Some of these particle-antiparticle pairs, e.g., the virtual electron-positron pairs, turn out to be charged. Such charged pairs act as an electric dipole. In the presence of an electric field, e.g., the electromagnetic field around an electron, these particle-antiparticle pairs reposition themselves, thus partially counteracting the field (yielding a partial screening effect, or a dielectric effect). The field therefore will be weaker than would be expected when the vacuum would be completely empty. The reorientation of the short-lived particle-antiparticle pairs is referred to as *vacuum polarization*.

3. In the everyday world, energy is always unalterably fixed; the law of energy conservation is a cornerstone of classical physics. But in the quantum microscopic world, energy can appear and disappear out of nowhere, in a spontaneous and unpredictable fashion. The uncertainty principle implies that particles can come into existence for short periods of time even when there is not enough energy to create them. In effect, they are created from uncertainties in energy. Since these particles do not have a permanent existence, they are called *virtual*

particles. Even though we can't see them, we know that these virtual particles are "really there" in empty space, because they leave a detectable trace of their activities. One effect of virtual photons, for example, is to produce a tiny shift in the energy levels of atoms. They also cause an equal change in the magnetic moment of electrons.

4. In modern physics, there is no such thing as "nothing." Even in a perfect vacuum, pairs of virtual particles are constantly being created and destroyed. And the existence of these particles is no mathematical fiction. Though they cannot be directly observed, the effects they create are quite real. A still more remarkable possibility is the creation of matter from a state of zero energy. The possibility arises because energy can be both positive and negative. The energy of motion or of mass is always positive, but the energy of attraction, such as that due to certain types of gravitational or electromagnetic fields, is always negative. Consider just the gravitational case, where the situation is especially interesting, for the gravitational field is no more than a warp-curved space. The energy locked up in a space-warp can be converted into particles of matter and antimatter. This occurs, for example, near a black hole. Thus, matter appears spontaneously out of empty space. Maybe the universe itself sprang into existence out of nothingness—a gigantic vacuum fluctuation. At least to some significant extent, the laws of contemporary physics allow for this possibility.

5. Many researchers see the vacuum as a central ingredient of twenty-first-century physics. We know now that the vacuum can have all sorts of wonderful effects over an enormous range of scales, from the submicroscopic to the cosmic. The vacuum's amazing properties all stem from a combination of what we call quantum theory and relativity. The uncertainty of the subatomic world manifests itself in random, causeless fluctuations in energy: the greater a fluctuation, the more fleeting its existence. Thanks to the famous equation $E = mc^2$, Heisenberg's uncertainty principle implies that particles can flit into and out of existence, their duration dictated only by their mass. This leads to the astonishing realization that, all around us, "virtual" subatomic particles are perpetually popping up out of nothing, and then disappearing again within about 10^{23} seconds. "Empty space" is thus not empty at all, but a seething sea of activity and processes that pervades the entire universe.

The Difference between the Causality/Determinism of Classical Physics and that of Quantum Physics

Heisenberg's "uncertainty relation" has had profound implications for such fundamental notions as causality and the determination of the future behavior of a subatomic particle, as well as philosophical implications for our conceptions of physical reality. Indeed, the uncertainty principle deeply questions the basic principles of classical physics. It states that for a pair of conjugate variables such as position/momentum or energy/time, it is impossible to establish a precisely determined value of each member of the pair at the same point in time. In particular, when the mass is very small (such as in subatomic objects), the uncertainties and thus the quantum effect become important, and tenets of classical physics are no longer applicable.

There is another important aspect, related to the preceding, that demonstrates clearly the difference between the classical and quantum visions of the physical world. In contrast to classical mechanics, the quantum vacuum is thought of as a seething froth of real particle-virtual particle pairs coming into and out of existence continuously and rapidly. Each of these strange pairs consists of a particle and its antiparticle, one of which has a *negative* energy and is thus called "virtual." Out of a singularity in space, which by definition is really is nothing, the pair simply comes into existence. Why? This is because the probability exists, because the universe is open to many (maybe infinite) possibilities that had not been predicted before, but yet could occur in the decades or millennia to come. The future, in other words, seems to be intrinsically unpredictable, at least according to quantum mechanics, but also to chaos theory and other system-dynamics theories. It is impossible to predict in a deterministic manner, from any information available a priori, which course a certain physical entity or event will take, or which it will not take. This means that quantum physics has, in a way, given the answer, if the original purpose was to know enough that given the circumstance we can predict (with precision and certainty) what will happen next.

As we have already stressed, this likely intrinsic, indeterminate, and in some sense fuzzy character of the quantum laws of nature (at the Planck scale) represents a strong denial of the philosophical idea associated with the deterministic vision of natural phenomena, according to which it is necessary for the very existence of science that the same conditions always produce the same results. And it is not merely our ignorance of the internal mechanisms, of the internal complications, that makes nature appear to contain both uncertainty and a host of possibilities. This seems somehow to be intrinsic. Nature herself does not know which way its constituents are likely to go.

Conceptually, quantum physics and geometry differ very deeply from classical physics and geometry. Besides the uncertainty principle, the implication of quantum mechanics for such notions as the nonlocality, entanglement, and identity of a physical entity is that it can play great havoc with classical intuition and our commonsense knowledge of the physical world. Generally, quantum mechanics assumes a fixed background space-time from which to define states on spacelike surfaces, as well as the unitary evolution between them. Quantum theory has changed as our conceptions of space and time have evolved. But quantum mechanics needs to be generalized still further for quantum gravity, a domain of where space-time geometry is fluctuating and lacks definite value.

The Similarities between the "Quantum Vacuum" and Plato's "Chora" (Space)

According to Plato, the *chora* is a space that is very hard to perceive: it is a continuous fluctuation, and it's the origin of all potentialities as well. So there is a striking resemblance between Plato's notion of space and the concept of "vacuum" in quantum physics. One might say that, to some extent the conception developed by quantum electrodynamics (QED) is a relativistic quantum field theory of classical electrodynamics, which describes how electrically charged particles—electrons, positrons, photons—interact by means of the exchange of photons. More precisely, QED can be described as a perturbative theory of the electromagnetic quantum vacuum. As such, it constitutes a theoretical and physical realization of Plato's idea (later interpreted by Aristotle to be the material substrate).

Let me offer some futher details. First, I wish to draw attention to the *chora*'s quality of being in motion (a potentiality, as we noted); as a cause of motion, it is also the source of time and change for the world of Becoming. Moreover, there is some kind of endless generativity of the *chora*, which is not restricted to its actual content, but is also a site of, say, virtual productivity. The *chora* in Plato's *Timaeus* is conceived as a space existing and acting between Being and Becoming. *Timaeus* dialogue indicates a return to a picture of the cosmic space that gave rise spontaneously to a world of becoming and change. Although the cosmos is always moving and changing, any specific change remains, strictly speaking, unaccomplished, unrealizable, and unhypothesizable. Of course, we should recall, despite what I just said, that the Greek philosophers did not like to admit the existence of a vacuum, asking themselves "how can 'nothing' be something?" which means that they point forward to a sort of logical (semantic) contradiction in the hypothesis of the physical reality of the vacuum. Plato found the idea of a vacuum inconceivable, essentially because he believed that all physical things were instantiations of an abstract ideal, and he could not therefore conceive of an "ideal" form of a vacuum. Similarly, Aristotle considered the creation of a vacuum impossible—nothing could not be something. Later Greek philosophers thought that a vacuum could exist outside the cosmos, but not within it. But this logical (semantic) objection fails when one considers the second aspect of my treatment of the question.

According to quantum mechanics and especially quantum electrodynamics, one must make a clear distinction between what we would intuitively understand to be an *absolute* void and what we experience as the vacuum of space. Of course, even in the deepest reaches of space we will find an atom or molecule here and there, and photons of energy are flying through at the speed of light continuously. But there is also a quantum potential that exists at every point in the vacuum of our three-dimensional physical space. Under the proper conditions, matter and energy can literally be made to materialize out of what we used to think of as nothing. So, according to the present mathematical models of quantum physical theories, it is *scientifically possible* (under certain physical conditions, energy, temperature, and so on) to create forms of matter and energy out of the vacuum.

The classical notion of empty space has obviously been updated. To quote the physicist John A. Wheeler, "No point is more central than this, that empty space is not empty. It is perhaps the seat of the most rich and surprising physics" (1968). There is, for instance, a remarkable possibility, which is the creation of matter from a state of zero energy. The possibility arises because energy can be both positive and negative. The energy of motion or the energy of mass is always positive, but the energy of attraction, such as that due to certain types of gravitational or electromagnetic fields, is negative. We have to consider just the gravitational case, where the situation is especially interesting, for the gravitational field is only a warp-curved space. The energy locked up in a space-warp can be concentrated into particles of matter and antimatter. This occurs, for example, near a black hole. Thus, matter appears spontaneously out of empty space—and here we actually join again Plato's philosophical idea of the *chora* and the cosmic space. Maybe the universe itself sprang into existence out of nothingness—a gigantic vacuum fluctuation. At least to some significant extent, the laws of contemporary physics allow for this possibility.

Remarks on the Quantum Effects
in Supersymmetric Quantum Field Theories

It is well known that there are quantum effects that are essential for explaining experimental data. Some of these quantum effects turn out to have infinite values and are
treated through "renormalization" techniques. But there are other quantum field theories,
called supersymmetric, where such dramatic effects are not present. Supersymmetry,
even if valid, must be a broken symmetry, and the quantum effects may again become
very large. This is an active area of theoretical physics at present.

The questions of quantum effects, renormalization, and broken symmetry are among
the most fundamental in theoretical physics today. Quantum effects were for example
considered in the development of cosmology. The properties of matter, radiation, spectral
lines, light scattering, and the statistics of Bose or Fermi—all these topics were taken into
account in the calculation of pressure, energy density, spectrum transport coefficients, and
so forth. Therefore, the righthand side of Einstein's general relativity equations, one might
say, already included quantum effects. Speaking of these effects now, we emphasize the
influence of space-time curvature on particles and fields, rather than the usual physics
of (flat) Minkowskian space. The most interesting effect is the creation of particles by
the gravitational field in vacuum. The reactions of type $e^+ + e^- = g + g$ (g being gravitons)
were considered and calculated in the 1930s and 1940s. Taking many coherent gravitons, one
obtains a classical gravitational wave. The creation of e^+e^- pairs obviously occurs in colliding
beams of (classical) gravitational wave, i.e., in a vacuum with time-dependent metric.

In a cosmological context, the excitation of fields, i.e., the creation of field quanta (for
example photons) in an expanding universe, was considered notably by Utiyama and
DeWitt in 1962. The general principle is that particle creation is due to the non-adiabatic
behavior of the corresponding field in a changing metric. The most important result of
the marriage between quantum theory (quantum effects) and general relativity (gravity)
is perhaps Hawking's theory (1975) of black hole evaporation. Concerning its physical
content, one can say, in a simplified manner, that collapse to a black hole is an event that
does not end in a static situation (unlike the collapse of a white dwarf to a neutron star,
for example).

Another important aspect concerns the relationship between conformal invariance
and spontaneous symmetry breaking, a topic so beautifully elucidated by Sidney Coleman in the 1980s. We recall first that for a relativistic field theory, the vacuum is Poincaré
invariant. Poincaré invariance implies that only scalar combinations of field operators
have nonvanishing vacuum expectation values (VEVs). The VEV may break some of the

internal symmetries of the Lagrangian of the field theory. In this case, the vacuum (that is, the ground state of the QFT) has less symmetry than the theory allows, and one says that spontaneous symmetry breaking has occurred. This means that, when the Hamiltonian (or the Lagrangian) of a system has some symmetry, but the ground state (i.e., the vacuum) does not, then one can say that spontaneous symmetry breaking (SSB) has taken place. When a continuous symmetry is spontaneously broken, massless gauge bosons (messenger particles that mediate the interactions) appear, corresponding to the remaining symmetry. This is called the Goldstone phenomenon, and the bosons are called Goldstone bosons.

In the electroweak model, the Higgs field acts as the order parameter breaking the electroweak gauge symmetry to the electromagnetic gauge symmetry. In empty space, the amplitude of the Higgs field differs from zero. This is also known as a "nonzero vacuum expectation value," the existence of which plays a fundamental role: it contributes mass to every elementary particle that has mass, including the Higgs boson itself. In particular, the acquisition of a nonzero vacuum expectation value spontaneously breaks electroweak gauge symmetry (often referred as the Higgs mechanism).

In quantum field theory (QFT) and the statistical mechanics of fields, renormalization refers to a collection of techniques used to construct mathematical relationships (or approximate relationships) between observable quantities, when the standard assumption, that the parameters of the theory are finite, breaks down, giving the result that many observables are infinite. Renormalization arose in quantum electrodynamics as a means of making sense of the infinite results of various calculations, and of extracting finite answers to properly posed physical questions. As pointed out by Gabriele Veneziano (1998), in Kaluza-Klein theory, in particular, both gauge theory and gravity diverge in the ultraviolet, in a similar way, another expected consequence of Kaluza-Klein unification; in other words, they become nonrenormalizable. On the one hand, quantum mechanics is essential to the success of the Kaluza-Klein idea (i.e., to unify electromagnetic and gravitational interactions by adding a new dimensional space, a circle, which allows for the quantization of momentum, hence of electric charge). At the same time, quantum field theory gives meaningless infinities and spoils the nice semiclassical results.

If the beautiful Kaluza-Klein idea is to be saved, we need quantum theory better than quantum field theory. But such theory already exists: it is called *superstring theory*. This theory relies crucially on the two ideas of supersymmetry and a space-time structure of eleven dimensions. Supersymmetry is a conjectured symmetry between fermions (particles having a half-integer spin, measured in quantum units) and bosons (particles with the same mass but integer spin). It is an inherently quantum mechanics symmetry, since the very concept of fermions is quantum mechanical. Bosonic quantities can be described by ordinary (commuting) numbers or by operators obeying commutation relations. Fermionic quantities involve anticommuting numbers or operators. Supersymmetry is an updating of special relativity to include fermionic as well as bosonic symmetries of space-time. In developing relativity, Einstein assumed that the space-time coordinates were bosonic; fermions had not yet been discovered! In supersymmetry, the structure of space-time is enriched by the presence of fermionic as well as bosonic coordinates. If true, supersymmetry explains why fermions exist in nature at all. Supersymmetry demands their existence. From experiments, we have some hints that nature may be supersymmetric.

But as we know, one of the most remarkable discoveries of recent decades is the theory worked out by Wess and Zunino in the 1970s that predicts a boson-fermion mass degeneracy that is not observed in nature, and thus the supersymmetry must be broken. The Goldstone fermions associated with spontaneous symmetry breaking possess the wrong property to be neutrinos, and hence the symmetry needs to be implemented as a local gauge invariance with the Higgs-Kibble mechanism in action.

In the original version of the standard model, the key to electroweak symmetry breaking is an entity called the Higgs particle. At high temperatures, Higgs particles, like other particles, move at random. But as the universe cools, Higgs particles combine into a "Bose condensate," an ordered state in which many particles share the same quantum wave function, leading—in the case of helium—to superfluidity. The electroweak symmetry is broken by the preferred "direction" of the Bose condensate. Although this proposal is simple and fits the known facts, it is unlikely to be the whole story. A seemingly artificial adjustment of parameters is needed to ensure that the Higgs particle's mass is small enough for the model to work. Numerous proposed alternatives solve this particular problem, but introduce puzzles of their own. One idea, motivated by a phenomenon that occurs in superconductors, is that the Higgs particle arises as a bound state. This would solve the problem of getting its mass right, but also requires a host of new particles and forces, none of which has yet been observed.

As we have seen above, a more radical and more convincing idea is "supersymmetry," a new symmetry structure of elementary particles in which quantum variables are incorporated into the structure of space-time. Other ideas about electroweak symmetry breaking go even further afield. One line of thought links this problem to extra dimensions of space-time, subnuclear in size, but observable at accelerators.

Finally, another line of thought links electroweak symmetry breaking to the dark energy of the universe, which astronomers have discovered in the past few years by observing that the expansion of the universe is accelerating. Although these theoretical proposals are all conceptually incomplete and experimentally unverified so far, the diversity and scope of ideas on electroweak symmetry breaking suggests that the solution to this riddle will provide new and significant insights into the particle physics search for the understanding of the laws of nature.

Cosmology is currently enjoying the most exciting period of discovery ever. Over the past ten years, a new, improved standard cosmology has emerged. It incorporates the highly successful standard hot Big-Bang cosmology, and extends our understanding of the universe to times as early as 10^{-32} sec, when the largest structures in the universe were still subatomic quantum fluctuations. The "New Cosmology" is characterized by: (i) a spatially flat, accelerating universe, (ii) an early period of rapid expansion (inflation), (iii) density inhomogeneities produced from quantum fluctuations during inflation, (iv) a composition of 2/3 dark energy, 1/3 dark matter, 1/200 h bright stars, (v) and a matter content of 29 ± 4% cold dark matter; 4 ± 1 % baryons; and ~ 0.3% neutrinos.

The "New Cosmology" is certainly not as well established as the standard hot big bang, but the evidence is mounting. One of its most striking features is the fact that

99.5% of the material in the Universe is dark, i.e., not manifested in the form of stars. By now, most scientists are familiar with dark matter, the name given by Zwicky to the un-detected matter whose gravity holds together cosmic structures from galaxies to the great clusters of galaxies. It is currently believed that the bulk of the dark matter exists in a sea of slowly moving elementary particles (CDM, "cold dark matter") left over from the earliest moments. The two leading candidates for the CDM particle are the axion and the neutralino. At present, there is no experimental evidence for the existence of either.

The CDM hypothesis is remarkable: it modestly holds that a new form of matter accounts for the bulk of the matter in the Universe. The hypothesis is being tested by experiments that seek to detect directly the dark matter particles that hold our own galaxy together, and by accelerator experiments that seek to produce neutralinos, whose mass is expected to be some 100 times that of the proton. Dark energy makes dark matter seem absolutely mundane! Dark energy is the term for the causative agent of the current epoch of accelerated expansion. According to the second Friedmann equation

$$\frac{\ddot{R}}{R} = -\frac{4\pi G}{3}\rho + 3p.$$

This stuff must have negative pressure, with magnitude comparable to its energy density, in order to produce accelerated expansion [recall that $q = -(\ddot{R}/R)/H^2$; R is the cosmic scale factor]. Further, since this mysterious stuff does not show its presence in galaxies or clusters of galaxies, it must be relatively smoothly distributed. That being said, dark energy has the following defining properties: (1) it emits light; (2) it has large, negative pressure, $p_X \sim -\rho_X$; and (3) it is approximately homogeneous (more precisely, it does not cluster significantly with matter on scales at least as large as clusters of galaxies). Because its pressure is comparable in magnitude to its energy density, it is more "energy-like" than "matter-like" (matter being characterized by $p \ll \rho$). Dark energy, nonetheless, is qualitatively quite different from dark matter.

It is well known that Einstein introduced the cosmological constant to balance the attractive gravity of matter. He quickly discarded the cosmological constant when the expansion of the Universe was discovered. Whether or not Einstein appreciated that his theory had predicted the possibility of repulsive gravity is unclear. The advent of quantum field theory made consideration of the cosmological constant obligatory, not optional: The only possible covariant form for the energy of the (quantum) vacuum,

$$T_{VAC}^{\mu\nu} = \rho_{VAC}g^{\mu\nu},$$

is mathematically equivalent to the cosmological constant. It takes the form of a perfect fluid with energy density ρ_{VAC} and isotropic pressure $p_{VAC} = -\rho_{VAC}$ (i.e., $w = -1$) and is precisely spatially uniform. Vacuum energy is almost the perfect candidate for dark energy. But here is the rub: the contributions of well-understood physics (say, up to the 100 GeV scale) to the quantum-vacuum energy add up to 10^{55} times the presently understood critical density. (Put another way, if this were so, the Hubble time would be 10^{-10} sec, and the associated event horizon would be 3 cm!) This is the well-known cosmological-constant problem.

How "Fock Space" Can Help
to Represent the Vacuum Fluctuations
in Quantum Field Theory

A consideration of the special structure of Fock space points to a very important aspect of (axiomatic) quantum field theory. Let us first recall that the origin of the Fock space concept lies in physics. A construction made by the Russian physicist V. A. Fock in 1932 suggested the means of passing from states of single objects to states of collections of these objects. Actually, it offered an abstract formulation of Hermite expansion in $L^2(R)$ as well, for $n = \infty$. It was then finally realized that Fock space is rather an algebra generated by a given Hilbert space and unity, and provided with a scalar product fulfilling certain natural requirements. The Fock space is an algebraic system or Hilbert space used in quantum mechanics to describe quantum states with a variable or unknown number of particles.

More precisely, the Hilbert space describes the quantum states for a single particle, and to describe the quantum states with n particles, or superpositions of such states, one must use a larger Hilbert space, the Fock space, which contains states for unlimited and variable numbers of particles. Fock states are the natural basis of this space. If $|\Psi_i\rangle$ is a basis of H, then we can agree to denote the state with n_0 particles in state $|\Psi_0\rangle$, n_1 particles in state $|\Psi_1\rangle$, ..., n_k particles in state $|\Psi_k\rangle$ by $|n_0, n_1, \ldots, n_k\rangle_v$, with each n_i taking the value 0 or 1 for fermionic particles and 0, 1, 2, ... for bosonic particles. Such a state is called a Fock state. Since $|\Psi_i\rangle$ are understood as the steady states of the free field, i.e., a definite number of particles, a Fock state describes a collection of non-interacting particles in definite numbers. The most general pure state is the linear superposition of Fock states. Two operators of paramount importance are the creation and annihilation operators, which upon acting on a Fock state, respectively, add and remove a particle in the ascribed quantum state. These operators serve as a basis for more general operators acting on the Fock space.

One important mathematical point is that Fock space can be examined in several categories of vector spaces. For example, Fock space has a natural interpretation as a space of holomorphic functions. This suggests that these "nonlinear" functions, obtained via Fock space, are not merely continuous but analytic.

Let us be a little more precise about the relationship between what vacuum state and the corresponding algebraic-geometric construction mean. There appear to be alternative choices for the "vacuum state," and this issue of "alternative vacua" seems to have some considerable importance in modern QFT. In fact, the choice of vacuum state is a matter of importance comparable with (and complementary to) the choice of the algebra A generated by creation and annihilation operators (see above), the latter defining, in a sense, the dynamics of QFT. In the case of free electrons, the two vacua $|0\rangle$ (containing no particles and no antiparticles) and $|\Sigma\rangle$ (in which all the negative-energy particle states are filled) can be considered as being, in a sense, effectively equivalent, despite the fact that $|0\rangle$ and $|\Sigma\rangle$ give us different Hilbert spaces. As we have already said, quantum states in which there are unlimited and variable numbers of particles, and superpositions of such states, require a Fock space, i.e. an infinite-dimensional vector space. These states are obtained by acting on $|0\rangle$, with an arbitrary element of A, i.e., an expression in creation and annihilation operators (polynomials or power series).

Again, the space of such states is referred to as Fock space, and it can be thought of as what is called a direct sum of Hilbert spaces with increasing numbers of particles. The number of particles in a state may be unlimited, as is so with the coherent states, which are, in a certain well-defined sense, the most "classical-like" of the quantum states. These are states of the form $e^{\Xi}|0\rangle$, where Ξ is the field operator associated with the particular field configuration **F** (a free real Maxwell field). **F** is defined to be the sum of the creation and annihilation operators (not normalized) corresponding to the positive- and negative-frequency parts of **F**, respectively. In recent years, there have been many attempts at interpreting different puzzling aspects of quantized Yang-Mills gauge theories, such as electrodynamics and chromodynamics, in the framework of a (Minkowskian) Fock space, and the problem of vacuum fluctuation is one of the most prominent aspects attendant upon these attempts.

Is the Word "Vacuum" Suitable for What Is Happening in Modern Physics?

I believe it impossible today to write a paper on modern quantum physics without ever mentioning the concept of vacuum. In so saying, I am affirming that the concept of "vacuum" is nowadays an essential part of theoretical physics. Of course, I am not denying the fact that the use of the word can be confusing, just as the use of many other philosophical and scientific terms can be. But that depends on our capacity to compose a definition (or statement) of these terms that is, as far as possible, accurate. Besides, the fact that there are still several notions and entities in modern physics that possess, let us say, a certain conceptual indeterminacy and vagueness doesn't mean that they are confusing and meaningless.

The concept of vacuum is an instructive example of what I am trying to say, for its initial indefiniteness was actually a reflection of an unexpected mathematical and physical depth. We should recall that the concept of vacuum, after all, was introduced by the great physicist P. Dirac in the 1930s and then largely employed by many other outstanding physicists, such as R. Feynman, J. Schwinger, B. S. DeWitt, J. A. Wheeler, T. D. Lee, B. Zumino, S. Coleman, S. Hawking, A. Linde, A. Vilenkin, and many others. Quantum electrodynamics, one of the most important theories of twentieth-century physics, would have been inconceivable without applying the concepts of vacuum, vacuum polarization, vacuum states, vacuum fluctuations, and vacuum energy, all of which have been well-defined in the framework of quantum field theory. Furthermore, the Casimir effect and the Lamb shift can be explained physically only by means of the concept of vacuum, which is clearly a dynamical and effectual entity.

More recently, the concept of vacuum has played a fundamental role in supersymmetric quantum field theories, and especially in string theory. In this theory, the concept of *vacuum energy* is a key ingredient of what is called the "landscape of string theory." The minima of this multidimensional landscape correspond to the stable configurations of space-time, which are called stable *vacua*. All these different stable vacua of the string landscape come into existence at various locations in the universe, thus forming innumerable sub-universes. This result might solve one of the most important and long-standing problems in modern theoretical physics—one related to vacuum energy. To Einstein, what we now think of as vacuum energy was an arbitrary mathematical term—a "cosmological constant"—that could be added to his equation of general relativity to make it consistent with his conviction that the universe is static. To obtain a static universe, he proposed that this constant take a positive value, but he abandoned the idea after observations proved the universe to be expanding. With the advent of quantum field theory, empty space—the vacuum—became a busy place, full of virtual particles and fields popping in and out of existence, and each particle and field carrying some positive or negative energy.

Concerning the nonzero value of the vacuum energy, the general idea of string theory is that the complicated geometries of hidden dimensions might produce a spectrum for vacuum energy that includes values in the experimental window. If the string landscape picture is valid, a nonzero vacuum energy should be observed, most likely not much smaller than 10^{-118} Λp (the value of Λp is about 1094 grams per cubic centimeter, or one Planck mass per cubic Planck length). So there are two key features in this new picture: first, the problem that the change of the vacuum configuration is linked with the multi-dimensional structure of the string landscape; second, that the vacuum energy might have an important effect on how the universe evolves.

Many researchers see the vacuum as a central part of twenty-first-century physics. We now know that the vacuum can have all sorts of wonderful effects over an enormous range of scales, from the microscopic to the cosmic. Let me conclude now with a general remark, namely, that often in science (as in philosophy and art) it is through imagining new models, inventing new concepts, and coining new words that we can reach a wide and deep vision of things. Thus we can "see" unexpected things: we "see" things that are far from what we would guess, far from what we could have conceived. Our imagination (and the concept of vacuum belongs to scientific creativity) is stretched to the utmost, not, as in fiction, only to imagine things that are not really there, but to comprehend those things that are there or could be there.

Mathematical Concepts and Techniques

In this appendix, we gather together mathematical facts—taken from differential geometry and topology and from algebraic geometry—that we use throughout the book. We focus on basic definitions, theorems, and reasonings, but we give no proofs. Our purpose is essentially to summarize some of the mathematical ideas that have played an important role in the development of the physical theories have we tackled in the text.

DEGREES OF MAPS

This appendix develops one of the most important tools in topology: the degree of a map[1] $f: M \to N$, where M and N are compact n-manifolds, N is connected, and $\partial M = \partial N = \varnothing$. This degree (see below) is an integer if M and N are oriented, an integer modulo 2 otherwise. Intuitively, the degree is the number of times f wraps M around N. The precise definition requires the theories of approximation, regular values, and orientation. If f is C^1 and $y \in N$ is a regular value, then the degree of f is the number of points in $f^{-1}(y)$ at which Tf preserves orientation, minus the number of points at which Tf reverses orientation.

It turns out that the degree of f is the same for all maps homotopic to f. This has two important consequences: it makes the degree of any given map easy to compute, and it gives us a convenient method of distinguishing homotopy classes. Moreover, the degree is the *only* homotopy invariant for maps into S^n. The degree is actually a special case of a more general geometrical concept called the *intersection number*. If M and N are submanifolds of W of complementary dimensions, and M and N are in general position, their intersection number is the algebraic number of points in $M \cap N$, each counted with appropriate sign determined by orientations. By means of transversality theory, we can define intersection numbers of maps, $M, N \to W$; again, we obtain homotopy invariants. If W is an n-dimensional oriented vector bundle $\zeta \to M$, then the self-intersection number of the zero section is called the *Euler number* $X(\zeta)$. This is an important isomorphism invariant of bundles. The Euler number of TM is the *Euler characteristic* $\chi(M)$. We can compute $X(\zeta)$ by means of sections of ζ. This leads to the computation of $\chi(M)$ as the sum of the indices of zeros of a vector field on M. One can also use the Morse inequalities to recompute $\chi(M)$ as the alternating sum of the Betti numbers of M.

Recall that the Euclidean n-space $\mathbb{R}^n$, $n \geq 1$, has the *standard orientation* ω^n given by any basis whose coordinate matrix has positive determinant (and the orientation of $\mathbb{R}^0$ is the number $+1$). Every n-dimensional submanifold of $\mathbb{R}^n$ is also given this orientation.

If (M, ω), (N, θ) are oriented manifolds, the *product orientation* $\omega \times \theta$ for $M \times N$ assigns to $(x, y) \in M \times N$ the orientation $\omega_x \oplus \theta_y$ of $(M \times N)_{(x, y)} = M_x \oplus N_y$. Let (M, ω) be an oriented ∂-manifold and $f : \partial M \times [0, \infty] \to M$ be a collar. Then $T_{\partial M} f$ induces an isomorphism of the trivial bundle $M \times \mathbb{R}$ onto the normal bundle v of ∂M in M. The standard orientation of $\mathbb{R}$ orients each fiber of $M \times \mathbb{R}$; via $T_{\partial M} f$, this induces an orientation ι of v that does not depend on the collar. In other words, v is oriented by *inward*-pointing vectors tangent to M at ∂M. We now have orientations ω and ι of M and v. From the exact sequence of vector bundles $0 \to T_{(\partial M)} \to T_{\partial M} M \to v \to 0$, we define the *induced orientation* $\omega / \iota = \partial \omega$ of ∂M. Thus $(e_1, \ldots, e_{n-1})$ is an orienting basis for $(\partial \omega)_x$ so long as $(e_1, \ldots, e_{n-1}, e_n)$ is an orienting basis for ω_x and e_n points *into* M at $x \in \partial M$. Let be θ an orientation of M. We usually give $M \times I$ the product orientation $\omega = \theta \times \omega^1$, where ω^1 is the standard orientation of I. It follows that $\partial \omega | M \times 0 = \theta$, and $\partial \omega | M \times 1 = -\theta$.

We speak here of "the oriented manifold M," not meaning the orientation explicitly. In this case, ∂M and $M \times I$ are also oriented manifolds, as is any submanifold of M of the same dimension. If $-M$ denotes the manifold M with the opposite orientation, then $\partial(M \times I) = (M \times 0) \cup (-M \times 1)$ as oriented manifolds. The closed unit n-disk $D^{n+1} \subset \mathbb{R}^{n+1}$ has the standard orientation. Therefore, its boundary S^n inherits an orientation, also called "standard." It is easy to verify that the stereographic projection from the north pole $P = (0, \ldots, 0, 1) \in S^n$ is an orientation-preserving diffeomorphism $S^n - P \approx \mathbb{R}^n$. Thus if $(e_1, \ldots, e_n)$ is an orienting basis for $\mathbb{R}^n \subset \mathbb{R}^{n+1}$, an orienting basis for S^n at the south pole $-P$ is $(e_1, \ldots, e_n)$, while at the north pole $(e_1, \ldots, e_{n-1}, -e_n)$ is orienting.

Let $A : \mathbb{R}^m \to \mathbb{R}^m$ be the antipodal map $A(x) = -x$. Since $\operatorname{Det} A = (-1)^m$, it follows that A preserves orientation of $\mathbb{R}^m$ if and only if m is even. The antipodal map of $\mathbb{R}^{n+1}$ restricts to a diffeomorphism of D^{n+1}. Since the map clearly preserves orientation of the normal bundle of ∂D^{n+1}, it follows that $A : S^n \to S^n$ preserves orientation if and only if n is odd.

Lemma 1. *Let (W, ω) be an oriented ∂-manifold. Suppose $K \subset W$ is an embedded arc that is transverse to ∂W at its endpoints $u, v \in \partial W$. Let κ be an orientation of K, and consider the quotient orientation $\omega | \kappa$ of the algebraic normal bundle of K. Then*

$$\omega_u | \kappa_u = (\partial \omega)_u \Leftrightarrow \omega_v | \kappa_v = -(\partial \omega)_v.$$

Let (M, ω), (N, θ) be compact oriented manifolds of the same dimension, without boundaries. Assume N is connected. Let $f : M \to N$ be a C^1 map and $x \in M$ a regular point of f. Put $y = f(x)$. We say x has *positive type* if the isomorphism $T_x f : M_x \to N_y$ preserves orientation, that is, if it sends ω_x to θ_y. In this case, we write $\deg_x f = 1$. If $T_x f$ reverses orientation, then x has *negative type*, and we write $\deg_x f = -1$. We call $\deg_x f$ the *degree of f at x*. Suppose $y \in N$ is any regular value for f. Define the *degree of f over y* to be

$$\deg(f, y) = \sum_{x \in f^{-1}(y)} \deg_x f;$$

If $f^{-1}(y)$ is empty, $\deg(f, y) = 0$. To indicate orientations, we also write

$$\deg(f, y) = \deg(f, y; \omega, \theta).$$

Reversing ω or θ changes the sign of $\deg(f, y)$.

To interpret $\deg(f, y)$ geometrically, suppose that $f^{-1}(y)$ contains n points of positive type and m points of negative type, so that $\deg(f, y) = n - m$. From the inverse function theorem, we can find an open set $U \subset N$ about y and an open set $U(x) \subset M$ about each $x \in f^{-1}(y)$, such that f maps each $U(x)$ diffeomorphically onto U, preserving or reversing orientation according to the type of x. Thus $\deg(f, y)$ is the algebraic number of time f covers U. For example, let S^1 be the unit circle in the complex plane. Let $M = N = S^1$, and $\theta = \omega$. If $f: S^1 \to S^1$ is the map $f(z) = z^n$, then $\deg(f, z) = n$, provided $z \neq 1$ when $n = 0$. If M is not connected, but has components $M_1, \ldots, M_k$, note that

$$\deg f = \sum_i \deg(f|M_i).$$

Of course, each M_j is given the orientation $\omega|M_j$ induced by the inclusion $M_j \subsetneqq M$.

Lemma 2. *Let W be a compact, oriented manifold of dimension $n + 1$, N be a compact oriented n-manifold without boundary, and $h : W \to N$ a C^∞ map. Let $y \in N$ be a regular value for both h and $h|\partial M$. Then $\deg(h|\partial W, y) = 0$.*

Corollary 1. *Let (M, ω) and (N, θ) be compact, oriented n-manifolds, $\partial M = \partial N = \varnothing$. Let N be connected, and $f, g : M \to N$ homotopic C^∞ maps having a common regular value $y \in N$. Then $\deg(f, y) = \deg(g, y)$.*

Lemma 3. *Let M, N be compact oriented n-manifolds without boundaries, $n \geq 1$, with N connected. Let $y, z \in N$ be regular values for a C^∞ map $f: M \to N$. Then $\deg(f, y) = \deg(f, z)$.*

Lemma 4. *Let M, N be manifolds and $f: M \to N$ be a continuous map. Then f can be approximated by C^∞ maps homotopic to f.*

We are ready to define the *degree of a map*. Let M, N be oriented compact n-manifolds, $n \geq 1$, with N connected and $\partial M = \partial N = \varnothing$. The *degree* $\deg f$ of a continuous map $f: M \to N$ is defined to be $\deg(g, z)$, where $g: M \to N$ is a C^∞ map homotopic to f and $z \in N$ is a regular value for g. By Lemma 5, such a g exists, and $\deg f$ is independent of g and z by Lemmas 3 and 4. If M and N are not oriented, perhaps even non-orientable, a *mod 2 degree* of f: $M \to N$ is defined as follows. Again, let $z \in N$ be a regular value for a C^∞ map $g : M \to N$ homotopic to f. Let $\deg_2(g, z)$ denote the reduction modulo 2 of the number of points in $g^{-1}(z)$. Then $\deg_2(g, z)$ is independent of g and z. This follows from the mod 2 analogue of Lemma 2, the proof of which reduces to the fact that a compact 1-manifold has an even number of boundary points. We define $\deg_2(f) = \deg_2(g, z)$. The results proved up to now apply to degrees of continuous maps to yield:

Theorem 1. *Let M, N be compact n-manifolds without boundary, with N connected. (a) Homotopic maps $M \to N$ have the same degree if M, N are oriented, and the same mod 2 degree otherwise. (b) Let $M = \partial W$, W compact. Suppose a map $f: M \to N$ extends to W. Then $\deg f = 0$ if W and N are orientable, and $\deg_2 f = 0$ otherwise.*

The degree is a powerful tool in studying maps. For examples, if $\deg f$ (or $\deg_2 f$) *is nonzero, then f must be surjective. For if f is not surjective, it can be approximated by a* homotopic C^∞ map g, which is not surjective. If $y \in N - g(M)$, clearly $\deg(g, y) = 0$. Here is an application of degree theory to complex analysis; it has the fundamental theorem of algebra as corollary. Let $p(z)$, $q(z)$ be complex polynomials. The rational function $p(z)/q(z)$ extends to a C^∞ map $f: S^2 \to S^2$, where S^2 denotes the Riemann sphere (the compactification of the complex field $\mathbb{C}$ by ∞). Then: f *is either constant or surjective.* The key of the proof is the observation that $z \in S^2$ is a regular point if and only if the complex derivative f' $(x) \neq 0$, and in this case the real derivative $Df_z : \mathbb{R}^2 \to \mathbb{R}^2$ has *positive* determinant. If f is not constant, then f' is not identically 0; hence, there is a regular point z. By the inverse function theorem, there is an open set $U \subset S^2$ about z, containing only regular points, such that $f(U)$ is open. Let $w \in f(U)$ be a regular value. Then $f^{-1}(w)$ is non-empty. Since every point in $f^{-1}(w)$ has positive type, it follows that $\deg(f, w) = \deg f > 0$. Therefore, f is surjective. A famous application of degree theory is the so-called "hairy ball theorem": *every vector field on S^{2n} is zero somewhere:* more picturesquely, a hairy ball cannot be combed. To prove this, suppose σ is a vector field on S^k that is nowhere zero. A homotopy of S^k from the identity to the antipodal map is obtained by moving each $x \in S^k$ to $-x$ along the great semicircle in the direction $\sigma(x)$. The existence of such a homotopy implies that the antipodal map has degree $+1$ and so preserves orientation; therefore, k is odd. The question of zeros of a vector field, or more generally, of a section of a vector bundle, can be approached more systematically with the theory of Euler numbers.

Lemma 5. *Let W be an oriented $(n + 1)$-manifold and $K \subset W$ be a neat arc. Let $V \subset \partial W$ be a neighborhood of ∂K and $f = V \to N$ be a map to an oriented manifold N, $\partial N = \varnothing$. Let $y \in N$ be a regular value of f and assume $\partial K = f^{-1}(y)$. Finally, assume that f has opposite degrees at the two endpoints of K. Then there is a neighborhood $W_0 \subset W$ of K and a map $g : W_0 \to N$ such that: (a) $g = f$ on $W_0 \cap V$, (b) y is a regular value of g, and (c) $g^{-1}(y) = K$.*

We can now state a basic extension theorem:

Theorem 2. *Let W be a connected oriented compact ∂-manifold of dimension $n + 1$. Let $f : \partial W \to S^n$ be a continuous map. Then f extends to a map $W \to S^n$ if and only if $\deg f = 0$.*

An analogue of the previous theorem for non-orientable manifolds is:

Theorem 3. *Let W be a connected compact non-orientable ∂-manifold of dimension $n + 1 \geq 2$. A map $f: \partial W \to S^n$ extends to W if and only if $\deg_2 f = 0$.*

We can now classify maps of all compact n-manifolds into S^n. Let $\simeq$ denote the relation of homotopy.

Theorem 4. *Let M be a compact connected n-manifold, $n \geq 1$. Let $f, g : M \to S^n$ be continuous maps. (a) If M is oriented and $\partial M = \varnothing$, then $f \simeq g$ if and only if $\deg f = \deg g$; and there are maps of every degree $m \in \mathbb{Z}$. (b) If M is non-orientable and $\partial M = \varnothing$, then $f \simeq g$ only if $\deg_2 f = \deg_2 g$; and there are maps of every degree $m \in \mathbb{Z}_2$. (c) If $\partial M \neq \varnothing$, then $f \simeq g$.*

LINKING NUMBERS

Let C_1 and C_2 be two mutually disjoint smooth closed curves in Euclidean 3-space. Then $Lk(C_1, C_2)$, indicating how closely they are interlinked with each other, is a quantity that was first given by Gauss: Now let C_i be expressed by the parameters $x_i = x_i(t_i)$ $(i = 1, 2)$, where $x_i(t_i)$ are continuously differentiable. Then the quantity

$$Lk(C_1, C_2) = -\left(\frac{1}{4\pi}\right)\int_{C_1}\int_{C_2}\left(\frac{1}{|x_2 - x_1|^3}\right)\det\left(x^2 - x^1, \frac{dx_1}{dt_1}, \frac{dx_2}{dt_2}\right)dt_2 dt_1$$

is an integer called the linking number of C_1 and C_2.

More generally, let M^n be an n-dimensional oriented combinatorial manifold (or C^r manifold $(r \geq 1)$) and K and K^* be its cellular decompositions, such that K^* is dual to K. Let z_1^r and z_2^s $(r + s = n - 1)$ be boundaries belonging to the complex K and K^*. Suppose that C^{r+1} is any chain of K whose boundary is z_1^r. The intersection number $[C^{r+1}]\cdot[z_2^s]$ does not depend on the choice of such a chain C^{r+1}. We set $Lk(z_1^r, z_2^s) = [C^{r+1}]\cdot[z_2^s]$ and call it the *linking number* of z_1^r and z_2^s. The *linking number* $Lk(\tilde{z}_1^r, \tilde{z}_2^s)$ of singular boundaries $\tilde{z}_1^r, \tilde{z}_2^s (r + s = n - 1)$ of M^n is similarly defined by considering the approximations $\tilde{z}_1^r, \tilde{z}_2^s$ of $\tilde{z}_1^r, \tilde{z}_2^s$ as belonging to a suitable cellular decomposition K and its dual K^*. The number $Lk(\tilde{z}_1^r, \tilde{z}_2^s)$ is bilinear with respect to $\tilde{z}_1^r, \tilde{z}_2^s$, and we have $Lk(\tilde{z}_1^r, \tilde{z}_2^s)$ $= (-1)^{rs+1}Lk(\tilde{z}_1^r, \tilde{z}_2^s)$. In the example of 3-dimensional Euclidean space R^3 shown in the left half of the first drawing, we have $Lk(\tilde{z}_1^1, \tilde{z}_2^2) = 1$, while $Lk(\tilde{z}_1^1, \tilde{z}_2^2) = 2$, for the example

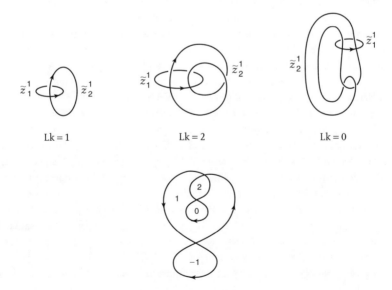

Lk = 1   Lk = 2   Lk = 0

The old $(\tilde{z}^1, 0)$ stays invariant as the point 0 moves in a connected component of the complement $R^2 - |\tilde{z}^1|$.

shown in the second drawing. In particular, if $\tilde{z}_1^r$ is homologous to 0 in $M^n - |\tilde{z}_2^s|$, then we have $\mathrm{Lk}(\tilde{z}_1^r, \tilde{z}_2^s) = 0$ in the third drawing. Generally, if $\tilde{z}_1^r$ and $\tilde{z}_1'^r$ are $\tilde{z}^s$ homologous in $M^n - |\tilde{z}_2^s|$, then $\mathrm{Lk}(\tilde{z}_1^r, \tilde{z}_2^s) = \mathrm{Lk}(\tilde{z}_1'^r, \tilde{z}_2'^s)$.

Let M^n be an n-dimensional oriented combinatorial manifold (or C^r manifold ($r \geq 1$)) with the nth Betti number $b_n = 0$, $\tilde{z}^{n-1}$ an $(n-1)$-dimensional singular boundary of M^n, and o a point of M^n that is not contained in $|\tilde{z}^{n-1}|$. We set $\mathrm{ord}(\tilde{z}^{n-1}, 0) = \mathrm{Lk}(\tilde{z}^{n-1}, 0)$ and call it the *order of the point* o with respect to $\tilde{z}^{n-1}$. For example, when $M^n = \mathbb{R}^2$ and $\tilde{z}^1 = \{f(t) | 0 \leq t \leq 1, f(0) = f(t)\}$, where f is a continuous function, the order $\mathrm{ord}(\tilde{z}^1, 0)$ is equal to the *rotation number* around o of a moving vector $of(t)$ as t varies from 0 to 1. This $\mathrm{ord}(\tilde{z}^1, 0)$ stays invariant as the point o moves in a connected component of the complement $\mathbb{R}^2 - |\tilde{z}^1|$ (fig. Q4). On the other hand, if $\tilde{z}_i^1 = \{f_i(t) | 0 \leq t \leq 1, f_i(0) = f_i(1)\} (i = 0, 1)$ are closed curves in $\mathbb{R}^2$ and the distance $\rho(f_0(t), f_1(t))$ is smaller than $\rho(f_0(t), o)$ for all t in the interval $[0, 1]$, then we have $\mathrm{ord}(\tilde{z}_0^1, o) = \mathrm{ord}(\tilde{z}_1^1, o)$ (Rouche's theorem).

COVERING SPACES AND MONODROMY

To explain what the *monodromy* is, we need first to introduce the concept of *covering spaces*. A continuous mapping $p : \tilde{N} \to N$ of an arcwise connected topological space $\tilde{N}$ onto a connected topological space N is called a *covering mapping* if the following condition (C) is satisfied: (C) Each point of N has an open neighborhood V such that every connected component of $p^{-1}(V)$ is mapped homeomorphically onto V by p. (Note that N is in fact arcwise connected.) If there is a covering mapping $p : \tilde{N} \to N$, we call $\tilde{N}$ a *covering space* of N and $(\tilde{N}, p, N)$ a *covering*. In particular, for a differentiable manifold N, if $\tilde{N}$ is also a differentiable manifold and p is differentiable, then $\tilde{N}$ is called a *covering (differentiable) manifold* of N. (In the theory of Riemann surfaces, a covering surface may have some branch points violating condition (C). (Upon removing such points, we obtain a covering space as defined above.) For each path $\tilde{a} : I \to N$ ($I = [0, 1]$) of N, a path $\tilde{a} : I \to \tilde{N}$ with $p \circ \tilde{a} = a$ is uniquely determined by the point $\tilde{a}(1) \in p^{-1}(a(1))$, and a bijection $a_\# : p^{-1}(a(1)) \to p^{-1}(a(0))$ is determined by $a_\#(\tilde{a}(1)) = \tilde{a}(0)$. Thus there exists a one-to-one correspondence between $p^{-1}(n)$ and $p^{-1}(n')$ for every pair of points a, a' of N, and $(\tilde{N}, p, N, p^{-1}(n_0))$ is a locally trivial fiber space with discrete fiber $p^{-1}(n_0)$. When the cardinal number of $p^{-1}(n)$ is a finite number m, we call $(\tilde{N}, p, N)$ an *m-fold covering*. In this case, for a loop $a(I, I) \to (N, n_0)$ with base point n_0, $a_\# : p^{-1}(n_0) \to p^{-1}(n_0)$ is a permutation of the m elements in $p^{-1}(n_0)$, and we obtain a homomorphism of the fundamental group $\pi_1(N) = \pi_1(N, n_0)$ of N into the symmetric group G_n, given by the correspondence $a \to a_\#$.

The permutation group $\mathcal{M}$, which is the image of this homeomorphism, is called the *monodromy group* of the m-fold covering. Two coverings $(\tilde{N}_i, p_i, N)$ ($i = 1, 2$) are said to be *equivalent* if there is a homeomorphism $\varphi : \tilde{N}_1 \to \tilde{N}_2$ with $p_2 \circ \varphi = p_1$; such a φ is called an *equivalence*. In particular, a self-equivalence $\varphi : \tilde{N}_1 \to \tilde{N}_2$ of a covering $(\tilde{N}, p, N)$ is called a *covering transformation*. The set π of all covering transformations forms a group by the composition of mappings, which is called the *covering transformation group* of $\tilde{N}$. We call $(\tilde{N}, p, N)$ a *regular covering* if, for each $n \in N$ and $\tilde{n}_1, \tilde{n}_2 \in p^{-1}(n)$, there exists a unique covering transformation that maps $\tilde{n}_1$ to $\tilde{n}_2$. In this case, the orbit space $\tilde{N}/\pi$ is homeomorphic to N, $(\tilde{N}, p, N, \pi)$ is a principal bundle, and the monodromy group $\mathcal{M}$ is isomorphic to π.

For a covering $(\tilde{N}, p, N)$, we call $\tilde{N}$ a *covering group* of N if $\tilde{N}$ and N are topological groups and p is a homeomorphism. Then $(\tilde{N}, p, N)$ is a regular covering, and its covering transformation group is isomorphic to $p^{-1}(e)$ (e is the identity element of N), which is a discrete subgroup lying in the center of $\tilde{N}$.

There are many examples of monodromy. Let's recall two very interesting examples. (1) Let S^1 be the unit circle in the complex plane with good cover $\mathsf{U} = \{U_0, U_1, U_2\}$. The map $\varphi : z \to z^2$ defines a fiber bundle $\pi : S^1 \to S^2$, each of whose fibers consists of two distinct points. Let $F = \{A, B\}$ be the fiber above the point 1. The cohomology $H^*(F)$ consists of all functions on $\{A, B\}$, i.e., $H^*(F) = \{(a, b) \in \mathbb{R}^2\}$. (2) The universal covering $\pi : \mathbb{R}^1 \to S^1$ given by $\pi(x) = e^{2\pi i x}$ is a fiber bundle with its fiber a countable set of points. The action of the loop downstairs on the homology H_0(fiber) is translated by $1 : x \to x + 1$. In cohomology, a loop downstairs sends the function on the fiber with support at x to the function with support at $x + 1$.

PATHS, LOOPS, AND HOMOTOPY

Recall that a *path* $\alpha : [0, 1] \to X$ is defined as a continuous map from the unit interval into a space X. The initial point is $\alpha(0) = x_0$ and the final point is $\alpha(1) = x_1$. A *loop* is a path where the initial and final points coincide, $\alpha(0) = \alpha(1) = x_0$, and x_0 is called the *base point*. As an example of a path, $\alpha : [0, 1] \to \mathbb{R}^2$ with $\alpha(s) = (\cos \pi s, \sin \pi s)$ is a path in the plane that traces out the upper half of the unit circle from $(1, 0)$ to $(-1, 0)$. Another example is the constant path, $c_x : [0, 1] \to X$ with $c_x(s) = x \ \forall \ s$. We can then define the properties of a *product* of paths, *the inverse* of a path, and, further, the fundamental concept of *a group of a set of paths in X* and the most essential properties of that group.

Let's now define *homotopy*. Let $\alpha, \beta : [0, 1] \to X$ be two loops with the same base point x_0. Then α and β are *homotopic*, written $\alpha \sim \beta$, if there exists a continuous map $F : [0, 1] \times [0, 1] \to X$ such that

$$F(s, 0) = \alpha(s) \quad F(s, 1) = \beta(s) \ \forall s \in [0, 1]$$
$$F(0, t) = F(1, t) = x_0 \ \forall s \in [0, 1].$$

F is called a homotopy between α and β. Let's unpack this formal definition. F is a function of two variables, s and t. The first variable, s, is the parameter that traces out the trajectory of a loop, so the first condition says that, for $t = 0$, F is the loop α, whereas, for $t = 1$, F is the loop β. The second variable, t, gives a continuous transition from $\alpha(s)$ to $\beta(s)$. The second condition tells us that every t corresponds to a loop in X, since the initial and final points are fixed to the base point x_0. Thus the homotopy $F(s, t)$ is a specific prescription for turning one loop into another in a continuous way, such that you always have a loop in X. This can be shown graphically by looking at the domain, which is a square in the s–t plane, and the image in X, shown in the figure below.

With the concept of homotopy, we can now turn our set of loops into a group. The homotopy class $[\alpha]$ of a loop α consists of all loops that are homotopic to α. The multiplication $*$ between paths naturally gives a multiplication rule for homotopy classes: $[\alpha] * [\beta] \equiv [\alpha * \beta]$. It is easy to show that this multiplication is well defined, which means

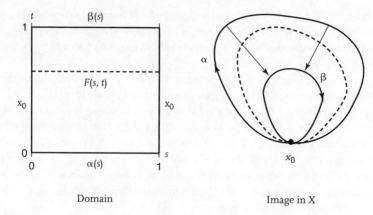

Domain   Image in X

Graphical representation of a homotopy between loops α and β, showing both the domain and the image in X.

that it doesn't depend on the representative loop in the homotopy class. In other words, if $\alpha \sim \alpha'$ and $\beta \sim \beta'$, we need to show $\alpha * \beta \sim \alpha' * \beta'$. The *fundamental group* or *first homotopy group* $\pi_1(X, x_0)$ is the set of homotopy classes $[\alpha]$ of loops in X with base point x_0, and with multiplication defined by $* : [\alpha] * [\beta] = [\alpha * \beta]$.

HOMOTOPY GROUPS OF SPHERES AND THE HOPF INVARIANT

The spheres S^n and their homotopy groups are basic objects in homotopy theory. Although much research has been done concerning these objects, there are still open problems. S^n is $(n-1)$-connected: $\pi_q(S^n) = 0$ $(q < n)$. According to an important theorem of W. Hurewicz[2] [see Milnor (1966)] and the homology of S^n, the homotopy groups of S^n in low dimensions are $\pi_q(S^n) = 0$ for $q < n$, and $\pi_n(S^n) = \mathbb{Z}$. The computation of $\pi_q(S^n)$ for $q \leq n$ is very long and complex (see Bott and Tu (1982) for a full account]. The first nontrivial computation of the homotopy of a sphere is $\pi_3(S^2)$. This can be done using the exact homotopy sequence of the *Hopf fibration*, as follows. Let S^3 be the unit sphere $\{(z_0, z_1) \mid |z_0|^2 + |z_1|^2 = 1\}$ in $\mathbb{C}^2$. Define an equivalence relation on S^3 by $(z_0, z_1) \sim (w_0, w_1)$ if and only if $(z_0, z_1) = (w_0, w_1)$ for some complex number λ of absolute value 1. The quotient $S^3/\sim$ is the complex projective space $\mathbb{C}P^1$ and the fibering

$$S^1 \to S^3$$
$$\downarrow$$
$$S^2 = \mathbb{C}P^1$$

is the *Hopf fibration*. From the exact homotopy sequence

$$\cdots \to \pi_q(S^1) \to \pi_q(S^3) \to \pi_q(S^2) \to \pi_{q-1}(S^1) \to \cdots$$

and the fact that $\pi_q(S^1) = 0$ for $q \geq 2$, we get $\pi_q(S^3) = \pi_q(S^2)$ for $q \geq 3$. In particular, $\pi(S^3) = \mathbb{Z}$. This homotopy group was first computed by H. Hopf in 1931 using a *linking number* argument that associates to each homotopy class of maps from S^3 to S^2 an integer now called the *Hopf invariant*. We give here an account of the Hopf invariant, first in the language of differentiable forms and then in terms of the linking numbers.

Let $f : S^3 \rightarrow S^2$ be a differentiable map, and let α be generator of $H^2_{DR}(S^2)$. Since $H^2_{DR}(S^3) = 0$, there exists a 1-form ω on S^3 such that $f*\alpha = d\omega$. As will be shown below, the expression

$$H(f) = \int_{S^3} \omega \wedge d\omega$$

is independent of the choice of ω. We define $H(f)$ to be the *Hopf invariant* of f. More generally, the same procedure defines the Hopf invariant for any differentiable map $f :$ $S^{2n-1} \rightarrow S^n$. If α is a generator of $H^n_{DR}(S^n)$, then $f*\alpha = d\omega$ for some $(n-1)$-form ω on S^{2n-1} and the Hopf invariant of f is

$$H(f) = \int_{S^{2n-1}} \omega \wedge d\omega.$$

Proposition (a) *The definition of the Hopf invariant is independent of the choice of ω.* (b) *For odd n, the Hopf invariant is 0.* (c) *Homotopic maps have the same Hopf invariant.*

Since homotopy groups can be computed using only smooth maps, it follows from Proposition 1(c) that the Hopf invariant gives a map

$$H = \pi_{2n-1}(S^n) \rightarrow \mathbb{R}$$

The map H is, in fact, a homeomorphism. Actually, the Hopf invariant is always an integer and is geometrically given by the *linking number* of the pre-images $A = f^{-1}(p)$ and $B = f^{-1}(q)$ of any two distinct regular values of f. In the classical case where $n = 2$, these two submanifolds are two "circles" embedded in S^3. To fix the ideas, we will first explain the linking concept for this case.

The linking number of two disjoint oriented circles A and B in S^3 can be defined in several quite different but equivalent ways. Choose a smooth surface D in S^3 with boundary A such that D intersects B transversally. Set the linking number to be

$$\mathrm{Lk}(A, B) = \Sigma_{D \cap B} \pm 1.$$

Here, the sum is extended over the points in the intersection of D with B, and the sign is given by the usual convention: at a point x in $D \cap B$, the sign is +1 or −1, according to whether the tangent space $T_x S^3$ has or does not have the direct sum orientation of $T_x D \oplus T_x B$. It must of course be shown that the linking number as defined is independent of the choice of D.

There are two other important definitions of the concept of the linking number: the *differential-form definition* and the *degree definition*. We will not deal here with these definitions.

We want only to show that the Hopf invariant of the Hopf fibration may also be found geometrically. First, recall the general definition. Let S^3 be the unit sphere in $\mathbb{C}^2$ and $f = S^3 \to \mathbb{C}P^1$ the natural map $f = (z_0, z_1) \to [z_0, z_1]$, where we write $[z_0, z_1]$ for the homogeneous coordinates on $\mathbb{C}P^1$. If $\mathbb{C}P^1$ is identified with the unit sphere S^2 in $\mathbb{R}^3$, say, via the stereographic projection, then the map $f: S^3 \to S^2$ is the *Hopf fibration*. To compute its Hopf invariant, one has to develop and prove the following five steps: (a) Find a volume form σ on the 2-sphere. (b) Write down a diffeomorphism $g : \mathbb{C}P^1 \xrightarrow{\sim} S^2$. (c) Pull the generator σ of $H^2(S^2)$ via g back to a generator α of $H^2(\mathbb{C}P^1)$. (d) Pull α back to S^3 via f and find a 1-form ω such that $f * \alpha = d\omega$ on S^3. (e) Compute $\int_{S^3} \omega \wedge d\omega$. Now, in order to interpret geometrically the Hopf invariant, one has to identify $S^3 - \{\text{north pole}\}$ with $\mathbb{R}^3$ via stereographic projection. This makes it possible for us to visualize the fibers of the Hopf fibration

$$S^1 \to S^3$$
$$\downarrow$$
$$S^2 = \mathbb{C}P^1$$

and to compute the linking number of the two fibers. We let $z_0 = x_1 + ix_2$, $z_1 = x_3 + ix_4$. Then the stereographic projection

$$p : S^3 - \{(0, 0, 0, 1)\} \mathbb{R}^3 = \{x_4 = 0\}$$

is given by

$$(x_1, x_2, x_3, x_4) \mapsto \left(\frac{x_1}{1 - x_4}, \frac{x_2}{1 - x_4}, \frac{x_3}{1 - x_4} \right).$$

This we see as follows. The line goes through the north pole $(0, 0, 0, 1)$, and the point (x_1, x_2, x_3, x_4) has parametric equation $(0, 0, 0, 1) + t(x_1, x_2, x_3, x_{4-1})$. Because it intersects $\mathbb{R}^3 = \{x^4 = 0\}$ at $t = 1/(1 - x_4)$, the intersection point is

$$\left(\frac{x_1}{1 - x_4}, \frac{x_2}{1 - x_4}, \frac{x_3}{1 - x_4}, 0 \right).$$

Note that the fiber S_∞ of the Hopf fibration over $[1, 0] \in \mathbb{C}P^1$ is $\{(z_0, 0) \in \mathbb{C}^2 | z_0| = 1\}$ and the fiber S_0 over $[0, 1]$ is $\{(0, 0, \cos\theta, \sin\theta) \in \mathbb{R}^4, 0 \leq \theta \leq 2\pi\}$, both oriented counterclockwise in their planes. So, via the stereographic projection, S_∞ corresponds to the unit circle in the (x_1, x_2)-plane, whereas S_0 corresponds to $\{(0, 0, \cos\theta/(1 - \sin\theta), 0 \leq \theta \leq 2\pi\}$, which is the x_3-axis with its usual orientation. Therefore, the linking number of S_∞ and S_0 is 1. By the geometric interpretation of the Hopf invariant as a linking number, the Hopf invariant of the Hopf fibration is 1.

Notes

1. The topological idea of the degree of a map is due to L. E. J. Brouwer (1912), who made fundamental contributions to the topology of manifolds. Further important results are due to S. Lefschetz and L. S. Pontryagin.

2. The so-called Hurewicz theorem (1941) is a basic result of algebraic geometry, connecting homotopy theory with homology theory via a group known as *Hurewicz homomorphism*. The Hurewicz theorem states that, if X is a path-connected space and $\pi_r(X) = 0$ for $r < n$, then $\pi_n(X) \cong H_n(X)(n \geq 2)$. There is also a corresponding version for relative homotopy and homology groups. [For the proofs, see Maunder (1980).]

Definition. The *Hurewicz homomorphism* $h_n : \pi_n(X) \to H_n(X)$ $(n \geq 1)$ is defined as follows. Let $\sigma_n \in H_n(S^n)$ be the standard generator; then if $[f] \in \pi_n(X)$ is represented by a map $f : S^n \to X$, define $h_n[f] = f*(\sigma_n)$. Clearly, this procedure is independent of the representative map chosen.

Theorem. *Let X be a path-connected space. Then $H_1(X)$ is the Abelianization of $\pi_1(X)$, i.e., if $[\pi_1(X), \pi_1(X)]$ is the commutator subgroup of $\pi_1(X)$, then $H_1(X) = \pi_1(X)/[\pi_1(X), \pi_1(X)]$.*

Theorem (Hurewicz Isomorphism Theorem). *Let X be a simply-connected path-connected CW complex. Then the first nontrivial homotopy and homology occur in the same dimension and are equal, i.e., given a positive integer $n \geq 2$, if $\pi_q(X) = 0$ for $1 \leq q < n$, then $H_q(X) = 0$ for $1 \leq q < n$ and $H_n(X) = \pi_n(X)$.*

Path Integral and Yang-Mills Connections

The purpose of this Appendix is to highlight some key features of the deep connection between geometrical structures and physical properties. More specifically, I want to stress that some geometric-differential structures, such as Yang-Mills connections, are at the very core of quantum field theories. Connections, curvature, bundles, and groups play a fundamental role in all recent unified theories of gauge fields. In quantum field theory, connections on principal bundles are called *gauge fields* and appear as generalizations of the electromagnetic field of Maxwell.

MATHEMATICAL INTRODUCTION TO PATH INTEGRALS

R. P. Feynman (1948) offered the solution of the Schrödinger equation as an integral of $e^{iL/h}$ over all possible paths $q(t)$, where $L = L(q, \dot{q})$ ($\dot{q} = (d/dt)q(t)$) is the classical Lagrangian for the Hamiltonian system. The integral is called the *Feynman path integral*. Consider the 1-body Schrödinger operator $H = H_0 + V$ (form sum), where V is the sum of a locally integrable function bounded below and a Kato perturbation on H_0. Let $b(t)$ ($t \geq 0$) be the Wiener process and $q(t) = hb(t)/(2m)^{1/2}$. For any L_2 functions f,

$$(G.1) \qquad (e^{-tH/h} f)(x) = E\left(f(x + q(t)) \exp\left\{ -\int_0^t V(x + q(s))\, ds/h \right\} \right)$$

for almost all x, where E denotes the expectation for the Wiener process. If V is a sum of L_2 and L_∞ functions (for spatial dimension ≤ 3), then the righthand side is continuous in x for $t > 0$. This is called the *Feynman-Kac formula*. Let L_0 be the Hamiltonian for a 1-dimensional harmonic oscillator with $m = \omega = h = 1$, and let ψ_0 be the eigenfunction $\psi_0(x) = \pi^{-1/4} \exp(-x^2/2)$. Consider $H = L_0 + V$ (form sum), where V is the sum of a locally integrable function bounded below and a Kato perturbation in L_0. Let $q(t)$ ($t \in \mathbf{R}$) be Gaussian random variables with mean 0 and covariance $E(q(t)q(s)) = 2^{-1} \exp(-|t - s|)$ (called the *oscillator process*). For any $f_i(x)$ in $L_2(\mathbf{R}, \psi_0^2 dx)$,

$$(G.2) \qquad \begin{aligned} &(j = 1,\dots,n) \text{ and } t_0 \leq t_1 \dots \leq t_n \leq t_{n+1}, (\psi_0, e^{-(t_1 - t_0)H} f_1 e^{-(t_2 - t_1)H} f_2 \dots f_n e^{-(t_{n+1} - t_n)H} \psi_0) \\ &= E\left(\left\{ \Pi_{j=1}^n f_j(q(t_j)) \right\} \exp\left\{ -\int_{t_0}^{t_{n+1}} V(q(s)) ds \right\} \right). \end{aligned}$$

The above path integral formulas are closely related to the *Trotter product formula*

(G.3) $$e^{-t(A+B)} = \lim_{n \to \infty} (e^{-tA/n} e^{-tB/n})^n (t \geq 0),$$

where A and B are self-adjoint operators bounded below and $A + B$ is essentially self-adjoint. (The same formula holds without the boundedness assumption when $t \in i\mathbf{R}$.)

THE PATH INTEGRAL APPROACH TO QUANTUM FIELD THEORY

The transition from quantum mechanics to quantum field theory is straightforward, but the underlying concept is a bit difficult to grasp. Basically, three issues must be dealt with. First, we demand that the theory be dynamically relativistic (in other words, $E = p^2/2m + V$ must be given the axe). Second, space and time variables must share equal billing. This is just another relativistic demand. In quantum mechanics, time is just a parameter, whereas position is an operator (that's why we see things like $|x\rangle$, whereas the object $|t\rangle$ is nonsensical). And third, quantum mechanics is primarily a one-particle theory, whereas a quantum field theory must somehow accommodate many particles (to account for particle creation and destruction). The path integral fulfills all of these requirements admirably.

Now here's the big leap in a nutshell: quantum field theory replaces the position coordinate x with a field $\phi(x)$, where the quantity x is now shorthand for x, y, z, t. That is, dimensional coordinates are downgraded from operators to parameters, just like t, so everything is on the same footing (in relativity, space and time are conjoined into *space-time*). This process of coordinate reassignment is known as *second quantization*. To reiterate (this is important), we must have a quantity whose functions are x, y, z, and t, something like the wave function $\Psi(\vec{x}, t)$. In quantum field theory, we assume the existence of a *quantum field* $\phi(x)$, which may also include specifications concerning particle spin, particle number, angular momentum, etc. In what is known as *canonical field theory*, the field itself is an operator (path integrals thankfully avoid this complication). If all of this makes sense to you (and even if it doesn't), then it shouldn't surprise you that

(G.4) $$Z = \int_{-\infty}^{\infty} \mathcal{D}\phi \exp\left[\frac{i}{h} \int_{-\infty}^{\infty} L(\phi, x^{\mu}, \partial_{\mu}\phi) \, d^4x \right],$$

where we assume that any and all coefficients (nasty or otherwise) are now lumped into the $\mathcal{D}\phi$ notation, which goes like (x_1) $(x_2) \ldots$ (Why it's called Z is just convention.) You might want to think of this quantity as the transition amplitude necessary for a field to propagate from the vacuum at $t = -\infty$ to the vacuum again at $t = \infty$, but I'm not sure that this prescription really describes it. A field can be just about anything, but you can look at it in this situation as a quantity that might describe a population of particles, energy fields, and/or force carriers at every point in space-time. And it is no longer appropriate to call (1) a path integral, since it does not describe the situation in terms of paths in space-time anymore. It is now called the *Z functional integral*.

You might now be wondering what the boundaries of the field are in terms of its possible values. Well, we can single out one very special field—the so-called vacuum state—in which the energy density of space-time in the vicinity of the system being considered is a minimum (usually zero), so that $Z \sim |0\rangle$. By this, we mean a state such that the modulus of the quantity Z cannot possibly assume any smaller value. By convention, we consider a vacuum state that arises at $t = \infty$. In propagator language, we say that $Z = \langle 0, \infty|0, -\infty\rangle$. Between these times, the field interacts with particles and other fields (and even creates them) in a manner prescribed by the Lagrangian. Thus, the field is born at $t = -\infty$, enjoys a "life" of some sort, and then dies at $t = \infty$ (that's why both integrals in (1) go from minus to plus infinity). Very simplistic, perhaps, but it seems to work well enough in practice. This choice of selecting the vacuum state as a starting point is fundamental to what is to follow. Because the path integral with interaction terms cannot be evaluated directly, a perturbative approach must be used. Selection of the vacuum or "ground" state ensures that the perturbation method will not "undershoot" the vacuum and give sub-vacuum results, which are meaningless. If the true vacuum state is not assigned from the beginning, then the system may jump to states of even lower energy. In QFT, this would be a disaster, because the method of solving for Z uses successive approximations (perturbation theory), and if we have a false vacuum, this methods fails utterly. In fact, the Higgs process[1] absolutely depends on fixing the true vacuum under a gauge transformation of the bosonic Lagrangian.

The form of the Lagrangian for a field depends on what kinds of particles and force carriers are going to be involved. Consequently, there are Lagrangians for scalar (spin zero) particles (also called *bosons*), spinors (spin 1/2 particles, also called *fermions*), and vector (spin one) particles. There's even a messy one of spin-2 gravitons. The simplest of these is the scalar or bosonic Lagrangian, and it is the one we will use here. The scalar Lagrangian for relativistic fields is given by

$$(G.5) \qquad L = 1/2\,[\partial_\mu \phi \partial^\mu \phi - m^2 \phi^2] - V.$$

(For a derivation, see any intermediate quantum mechanics text, for example, Sakurai, 1985.)

Just like the ordinary propagator in quantum mechanics, we're going to experience problems evaluating Z when the potential term V is not a linear or quadratic function of its arguments. The simplest interaction term for a scalar particle in quantum field theory turns out to be $V \sim \lambda \phi^4$, where λ is called a *coupling constant*. This gives rise to what is called a *self-interacting* field theory; that is, the field interacts with itself and with any particles that are created along the way. As a result, the integral for Z cannot be obtained in closed form, and we will have to resort to perturbation theory, as previously indicated. This leads to a very interesting interpretation of particle creation and propagation as a consequence of this model—at every order in the perturbative expansion (including zero order), particles appear and begin to propagate about the space-time stage. Since in principle there is an infinite number of space-time points where interactions can occur, the number of particles involved can also be infinite. However, the total number of all interactions is fixed by the number of λ that enter the perturbative expansion of Z. Thus,

(G.6) $Z = \int \mathcal{D}\phi \exp \{i/\hbar \int [1/2(\partial_\mu \phi \, \partial^\mu \phi - m^2\phi^2) - \lambda\phi^4] \, d^4x\}.$

But this integral cannot be attacked in this form. The main problem is the infinite-dimensional integral; it is simply too unwieldy. We will have to make some changes before a perturbative solution can be employed.

The expression (G.3) can be rewritten as

(G.7) $Z(J) = \int \mathcal{D}\phi \exp [i/2 \int [\partial_\mu \varphi \partial^\mu \varphi - m^2\varphi^2] \, d^4x\} - \iint J(x')\Delta_F(x'' - x') J(x'') d^4x' \, d^4x''].$

Because the $J(x)$ that appears under this integral is an explicit function of the space-time coordinates x, *and not a function of* φ, the J integral term *can be taken out of the infinite-dimensional integral altogether*:

(G.8) $Z(J) = \exp [i/2 \iint J(x')\Delta_F(x' - x'') J(x'') \, d^4x' \, d^4x'']$

     $\int \mathcal{D}\phi \exp [i/2 \int (\partial_\mu \varphi \partial^\mu \varphi - m^2\varphi^2)] \, d^4x.$

Here, the residual integral over $\mathcal{D}\phi$ is just some number; we call it N and we set $N = 1$, because this allows us to continue using normalized amplitude. We then have, finally,

(G.9) $Z(J) = \exp [- i/2 \int J(x)\Delta F (x - x') J(x') \, dx \, dx'],$

where I'm now using the residual integral sign and dx for brevity. This, in fact, is an enormous achievement, for we have successfully rid ourselves of that infinite-dimensional integral and replaced it with two four-dimensional integrals. From here on out, everything can be done to involve taking successive derivatives of Z with respect to the $J(x)$. This is the main reason why Z was "simplified" in this way—it provides a parameter, $J(x)$, with which the solution of $Z(\lambda)$ can be straightforwardly developed.

3. YANG-MILLS G-CONNECTION.

Let $P = (P, \pi, M, G)$ be a differentiable principal fiber bundle over a compact oriented Riemannian manifold M with group G. Then, fiber bundles $G_p = P \times_c G$ and $\mathfrak{g}_p = P \times_{\mathrm{Ad}} \mathfrak{g}$ associated with P are induced naturally from the group conjugation $c : G \to \mathrm{Aut}(G)$ and the adjoint representation $\mathrm{Ad} : G \to \mathrm{Aut}(\mathfrak{g})$, respectively. A (local) section of G_p is called a (local) *gauge transformation* of P. The set of all global transformations, which is denoted by G_p, has a group structure.

A locally faithful representation ρ of G to an n-dimensional complex vector space F with a fixed basis $(\zeta_1, \ldots, \zeta_n)$ defines a differentiable complex vector bundle $E = P \times_\rho F$ associated with P. Every point x of P is identified with a linear mapping $\bar{x}$ of F onto the fiber $E_{\pi(x)}$ defined by $\bar{x} : \zeta_J \to e_i$, where e_i denotes the equivalence class of $\{x, \zeta_i\} \in P \times F$, $1 \leq I \leq n$. In a manner similar to the case of an affine connection, a connection in P with connection form ω gives a notion of parallel displacement of E, as follows: Let $c = p_t$ $(0 \leq t \leq 1)$ be a curve in M and $c^* = x_t$ be a lift of c to P. The mapping $\bar{x}_t^0 \bar{x}_0^{-1} : E_{P_0} \to E_{P_t}$ is called the *parallel displacement* of E_{p_0} onto E_{p_t} along c. Let X be a vector field on M, and let φ be a differentiable section of E. The covariant derivative $\nabla_X\varphi$ of φ in the direction of X is defined as follows: Let p_0 be a point of M, $c = p_t \ (-\varepsilon \le t \le \varepsilon)$ be an integral curve of X through p_0, and $c^* = x_t$ be a lift of c to P. We set

$$(G.10) \qquad (\nabla_X\varphi)_{P_0} = \lim_{t\to 0}\frac{1}{t}(\bar{x}_t \circ \bar{x}_0^{-1}\,\varphi_{P_t}) = \varphi_{P_0}).$$

Then ∇_X is also a differentiable section of E.

The mapping $(X, \varphi) \to \nabla_X\varphi$ satisfies the following conditions: (i) $\nabla_X\varphi$ is linear with respect to X and φ; (ii) $\nabla_{fX}\varphi := f\nabla_X\varphi$; and (iii) $\nabla_X(f\varphi) = (Xf)\varphi + f\nabla_X\varphi$, where f is a differentiable function on M. From these conditions, it is seen that the mapping $X \to \nabla_X\varphi$ for a fixed section φ of E defines a differential form of degree 1 with values in E, denoted by $\nabla\varphi$. ∇ is called a G-connection on E (induced from the connection in P). A linear operator $d^\nabla : \Gamma(\Lambda^p \otimes E) \to \Gamma(\Lambda^{p+1} \otimes E)$, defined by $d^\nabla(\alpha \otimes \varphi) = d\alpha \otimes \varphi + (-1)^p\,\alpha \wedge \nabla\varphi$ for a differential form α of degree p and a differentiable section φ of E is called a *covariant exterior differentiation*. Here, $\Gamma(\Lambda^p \otimes E)$ denotes the set of all differential forms of degree p with values in E.

The curvature form R^∇ of ∇ is defined by $R^\nabla(X,Y)\varphi = \nabla_X(\nabla_Y\varphi) - \nabla_Y(\nabla_X\varphi) - \nabla_{[X,Y]}\varphi$, where X and Y are vector fields on M and φ is a differentiable section of E. In terms of the curvature form Ω of the connection in P, it can be defined also by

$$(G.11) \qquad R_P^\nabla(X,Y)\varphi = \bar{x} \circ \Omega_x(X^*, Y^*) \cdot \bar{x}_0^{-1}(\varphi),$$

where X and Y are in T_pM and $x \in \pi^{-1}(p)$, and X^*, Y^* are lifts of X, Y to P, respectively, and φ is an element of E_p. Since Ω is a tensorial form of type ad and the differential ρ^* of ρ induces a faithful representation of $\mathfrak{g}$ to $\mathfrak{gl}(F)$, R^∇ can be regarded as a differential form of degree two with values in the bundle $\mathfrak{g}_p = P \times_{\mathrm{Ad}} \mathfrak{g}$. The curvature form R^∇ satisfies Bianchi's identity: $d^\nabla R^\nabla = 0$, where d^∇ is a covariant exterior differentiation with respect to the G-connection on $\mathfrak{g}_p$ canonically induced from the connection in P.

We denote by $\mathscr{G}$ the set of all G-connections on E. Now let G be a compact semi-simple Lie group. A functional $\mathscr{L} : \mathscr{G} \to \mathbf{R}$ defined by $\nabla \to \mathscr{L}(\nabla) = 1/2 \int_M \mathrm{tr}(R^\nabla \wedge *R^\nabla)$ is called the *action integral* (*Yang-Mills functional*) of ∇, where $*$ is the Hodge's star operator, given by the fixed orientation of M. The group $\mathscr{G}_p$ acts on $\mathscr{L}$ by $\nabla \to f^{-1} \circ \nabla \circ f$ for $f \circ \mathscr{G}_p$ and the curvature form is then transformed by this action as $R^\nabla \to \mathrm{Ad}(f^{-1})\,R^\nabla$. Thus $\mathscr{L}$ is $\mathscr{G}_p$-invariant.

A connection ∇ is called a *Yang-Mills G-connection* if ∇ is a critical point of $\mathscr{L}$. The Euler-Lagrange equation of $\mathscr{L}$ is given by $d^{\nabla^*}R^\nabla = 0$ by the aid of the formal adjoint operator d^{∇^*} of d^∇. This equation, called the *Yang-Mills equation*, is a system of nonlinear second-order elliptic partial differential equations. When M is a 4-dimensional vector space $\mathbf{R}^4$ with Minkowskian metric, and G is the Abelian group $U(1)$, the Yang-Mills equation coincides with Maxwell's equations for an electromagnetic field. Thus the Yang-Mills equation for a non-Abelian group G is a natural extension of Maxwell's equations.

In fact, the theory of Yang-Mills connections has its origin in the field theory of physics. In the case of dim $M = 4$, special Yang-Mills G-connections occur. A G-connection ∇ satisfying the condition $* R^\nabla = R^\nabla$ (resp. $* R^\nabla = - R^\nabla$) as a differential form of degree two is called a *self-dual* (resp. *anti-self-dual*) G-connection. From Bianchi's identity and the expression of $d^{\nabla*} : d^{\nabla*} = - * \circ d^\nabla \circ *$, it follows that every self-dual (anti-self-dual) G-connection gives a solution to the Yang-Mills equation. Since the first Pontryagin number $p_1(E)$ is given by $p_1(E) = -\dfrac{1}{4\pi^2} \int_M \mathrm{tr}(R^\nabla \wedge R^\nabla)$ by virtue of the Chern-Weil theorem, the action integral satisfies $\mathscr{L}(\nabla) \geq 2\pi^2 |p_1(E)|$ for every ∇ in $\mathscr{G}$, and the equality holds if and only if ∇ is self-dual or anti-self-dual.

Explicit forms have been obtained for (anti-)self-dual connections over the 4-sphere S^4 by many interesting methods. And it has been shown that moduli space of self-dual G-connections (i.e., the set of all solutions to the Yang-Mills equation modulo $\mathscr{G}_p$) over S^4 has the structure of a Hausdorff manifold of dimension $p_1(\mathfrak{g}_p) - \dim G$ for every principal bundle P with group G. It is not yet known whether there exists a Yang-Mills G-connection over S^4 whose holonomy group is an open subgroup of G and which is neither self-dual nor anti-self-dual. The following is one of the few facts concerning the properties of Yang-Mills G-connections: If a Yang-Mills G-connection, $G = SU(2)$, $SU(3)$, or $O(3)$, over S^4 is weakly stable, i.e., if the second variation of $\mathscr{L}$ is positive semidefinite, then it is self-dual or anti-self-dual.

Note

1. In particle physics, the Higgs mechanism is a theoretical framework that concerns the origin of the property "mass" of elementary particles; technically, it yields the only consistent explanation of how the masses of the W and Z bosons arise through spontaneous electroweak symmetry breaking. More generally, the Higgs mechanism is the process by which the gauge boson in any gauge theory acquires a nonzero mass. For other particles, e.g., for fermions, a Higgs mechanism can explain the masses, again in a gauge-invariant way. The simplest realization of the Higgs mechanism in the Standard Model requires an extra Higgs field that interacts with the gauge fields, and has a nonzero value in its lowest energy state, a vacuum expectation value. This means that all of space is filled with the background Higgs field, the so-called Higgs condensate. Interaction with this background field changes the low-energy spectrum of the gauge fields, and the gauge bosons become massive. The Higgs field has a nontrivial self-interaction, like the Mexican hat potential, which leads to spontaneous symmetry breaking: in the lowest energy state, the symmetry of the potential (which includes the gauge symmetry) is broken by the condensate. Analysis of small fluctuations of the fields near the minimum reveals that the gauge bosons and other particles become massive. In the Standard Model, the Higgs field is an $SU(2)$ doublet, a complex spinor with four real components, which is charged under the Standard Model symmetry group $U(1)$, which is the electromagnetism's gauge symmetry of charge -1. After symmetry breaking, three of the four degrees of freedom in the Higgs field mix with the W and Z bosons, while the one remaining degree of freedom becomes the Higgs boson—a new scalar particle. The Higgs mechanism in the Standard Model successfuly predicts the mass of the $W^\pm$ and Z weak gauge bosons, which are of course massless. If the Higgs mechanism were not there, these particles would acquire much

smaller masses from QCD quark condensates instead. In the Standard Model of particle physics, the same Higgs mechanism that breaks the Standard Model gauge group down to the subgroup $U(1)$ of electromagnetism is also responsible for giving all the leptons and quarks their masses. The fermions in the Standard Model are chiral, and different chiralities have different charges. The chiral fermions can come together in pairs to make massive fermions by absorbing Higgs bosons from the condensate.

Notes

CHAPTER 1. INTRODUCTION

1. In particle physics and physical cosmology, the Planck scale is an energy scale situated around 1.22×10^{19} GeV (which corresponds by the mass-energy equivalence to the Planck mass 2.17645×10^{-8} kg) at which quantum effects of gravity become strong. The term "Planck scale" can also refer to a length scale, which corresponds to $\ell_p = 1.616252 \times 10^{-35}$ meters, and to a time scale equivalent to 5.39121×10^{-44} s. At these scales, the description of subatomic particle interactions in terms of quantum field theory breaks down (due to the non-renormalizability of gravity). At the Planck scale, the strength of gravity is expected to become comparable to the other forces, and it is theorized that all of the fundamental forces are unified at that scale, but the exact mechanism of this unification remains unknown. The Planck length is related to Planck energy by Heisenberg's uncertainty principle. At this scale, the concepts of size and distance break down, as quantum indeterminacy, i.e., the apparently necessary incompleteness in the description of a physical system, becomes virtually absolute.

2. The theory of Hilbert space arose from problems in the theory of integral equations. D. Hilbert noticed (1924) that a linear integral equation can be transformed into an infinite system of linear equations for the Fourier coefficients of the unknown function. He considered the linear space ℓ_2, consisting of all sequences of numbers $\{x_n\}$ for which $\sum_{n=1}^{\infty} |x_n|^2$ is finite, and then defined, for each pair of elements $x = \{x_n\}, y = \{y_n\} \in \ell_2$, their inner product as $(x, y) = \sum_{n=1}^{\infty} x_n \bar{y}_n$. The space can be regarded as an infinite-dimensional extension of the notion of a Euclidean space. F. Riesz considered the space of functions now termed L_2-space and rendered a satisfactory answer to the Fourier expansion problem. In 1932 J. von Neumann established a rigorous foundation of quantum mechanics, employing Hilbert spaces and their spectral expansion of self-adjoint operators.

To define a Hilbert space, let K be the field of complex or real numbers, the elements of which we denote by $\alpha, \beta, \ldots$. Let H be a linear space over K, and to any pair of elements $x, y \in H$ let there correspond a number $(x, y) \in K$ satisfying the following five conditions: (i) $(x_1 + x_2, y) = (x_1, y) + (x_2, y)$; (ii) $(\alpha x, y) = \alpha(x, y)$; (iii) $(x, y) = \overline{(x, y)}$; (iv) $(x, x) \geq 0$; and (v) $(x, x) = 0 \Leftrightarrow x = 0$. We then call H a *pre-Hilbert space* and (x, y) the *inner product* of x and y.

With the norm $\|x\| = \sqrt{(x, x)}$, H is a normed linear space. If H is complete with respect to the distance $\|x - y\|$ (i.e., $\|x_n - x_m\| \to 0(m, n \to 0)$, implying the existence of $\lim x_n = x$), we then call H a *Hilbert space*. According as K is complex or real, we call H a *complex* or *real Hilbert space.* A Hilbert space is also a Banach space. A Banach space B is a normed space with an associated metric $d(x, y) = \|x - y\|$, such that every Cauchy sequence in B has a limit in B. A normed linear space with norm $\|x\|$ can be made a pre-Hilbert space, by defining an inner product (x, y) so that $\|x\| = \sqrt{(x, x)}$, if and only if the equality $\|x + y\|^2 + \|x - y\|2 = 2(\|x\|^2 + \|y\|^2)$ holds for any x, y.

In the mathematically rigorous formulation of quantum mechanics, developed by P. Dirac and J. von Neumann in the 1930s, the possible states (more precisely, the pure states) of a quantum mechanical system are represented by unit vectors (called *state vectors*) residing in a complex separable Hilbert space, known as the *state space*, well defined up to a complex number of norm 1 (the phase factor). In other words, the possible states are points in the projectivization of a Hilbert space, usually called the *complex projective space*. The exact nature of this Hilbert space is dependent on the system; for example, the position and momentum states for a nonrelativistic spin zero particle are the space of all square-integrable functions, whereas the states for the spin of a single proton are unit elements of the two-dimensional complex Hilbert space of spinors. Each observable is represented by a self-adjoint linear operator acting on the state space. Each eigenstate of an observable corresponds to an eigenvector of the operator, and the associated eigenvalue corresponds to the value of the observable in that eigenstate.

CHAPTER 2. THE ROLE OF VACUUM IN MODERN PHYSICS

1. This has particular relevance to unstable nuclei. The fact that such a nucleus (of, say, uranium) exists means that there is a limit to the time—namely the particle's lifetime—during which the particle's energy can be ascertained. Accordingly, Heisenberg's relation gives us a fundamental energy uncertainty, for an unstable particle or nucleus, that is reciprocally related to its lifetime. Because of Einstein's $E = mc^2$, this gives us a fundamental uncertainty in its *mass*, as well. The wave-function of an unstable particle deviates from its being of the stationary form $e^{-iEt/h}$ for some definite real energy value E, there being also an exponential decay factor to be considered. Because an energy eigenstate does not exist, there is a resulting spread in the measured energy.

2. This is not a statement about the inaccuracy of measurement instruments, nor a reflection on the quality of experimental methods; rather, it arises from the wave properties inherent in the quantum-mechanical description of nature. Even with perfect instruments and technique, the uncertainty is inherent in the nature of things.

3. Recall that, following the cosmological models developed by Alexander Friedmann (1922) and Georges Lemaître (1933), Howard Robertson and Arthur Walker introduced a metric that is an exact solution of Einstein's field equations of general relativity; their metric describes a simple, homogeneous, isotropic, expanding or contracting universe, and they assumed that the spatial component of the metric can be time-dependent.

4. Non-Abelian gauge theory is now applied to electroweak theory and quantum chromodynamics (QCD), not to strong isospin symmetry, as originally speculated by Yang and Mills. Historically, in 1954, attempting to resolve some of the great confusion in elementary particle physics, Chen Ning Yang and Robert Mills introduced non-Abelian gauge theories as models by which to understand the strong interactions holding to-

gether the nucleons in atomic nuclei. Generalizing the gauge invariance of electromagnetism, they attempted to construct a theory based on the action of the (non-Abelian) SU(2) symmetry group on the isospin doublet of protons and neutrons. This is similar to the action of the U(1) group on the isospin fields of quantum electrodynamics (QED). In particle physics, the emphasis was on using quantized gauge theories. This idea later found application in the quantum field theory of the weak force, and in its unification with electromagnetism in the electroweak theory. Gauge theories became even more attractive when it was realized that non-Abelian gauge theories reproduced a feature called *asymptotic freedom*. This prediction, first advanced in the early 1970s by David Politzer and by Frank Wilczek and David Gross, establishes that in very high-energy reactions, quarks and gluons interact very weakly. (The other peculiar property of QCD is the *confinement*, which means that the force between quarks does not diminish as they are separate; thus, it would take an infinite amount of energy to separate two quarks; they are forever bound into hadrons, such as the proton and the neutron.) Asymptotic freedom was believed to be an important characteristic of strong interactions. This motivated searching for a strong force gauge theory. This theory, now known as quantum chromodynamics, is a gauge theory whose SU(3) group acts on the color triplet of quarks.

5. There are several other interesting Yang-Mills theories. For example, it has been suggested that the standard model, based on the group SU(3) × SU(2) × U(1), is a subgroup of a larger simple group, such as SU(5). Theories of this kind, which attempt to unify interactions, are sometimes known as grand unified theories (GUTs). Another possible GUT is based on the group SU(10). These theories are somewhat speculative, and direct measurements are not possible today, since the unification energies are too high. But they often make indirect predictions, such as proton decay or the existence of magnetic monopoles. So far, there has been no solid experimental confirmation of GUTs.

CHAPTER 3. THE QUANTUM VACUUM IN THE EARLY UNIVERSE

1. Erwin Schrödinger proved that quantum wave functions coevolve with the curved space-time of the Friedmann universe. More precisely, he found that the plane-wave eigenfunctions characteristic of flat space-time are replaced in the curved space-time of the Friedmann universe by wave functions that are not precisely flat, and have wavelengths that are directly proportional to the Friedmann radius. This means that the eigenfunctions change wavelength as the radius of the universe changes, and hence the quantum systems they describe follow. In an expanding universe, quantum systems expand. In a contracting universe, they contract. Schrödinger's derivation explains the Hubble redshift of photons in an expanding universe, as well as the energy changes of moving particles, and establishes the coevolution of atoms and other quantum systems with space-time geometry. In other words, these changes in quantum systems may equivalently be viewed as a logical consequence of the fact that the energy and momentum of an "isolated" system can change in general relativity when the space-time geometry of the universe changes. Thus, the assumption often made that small quantum systems are isolated, and that their properties remain constant as the Friedmann universe evolves, is

incompatible with relativistic quantum mechanics, and with general relativity. This fact has, for example, a simpler interpretation in wave mechanics: all wavelengths, being inversely proportional to the momenta, simply expand with space.

CHAPTER 4. THE PROBLEM OF THE VACUUM AND THE CONCEPTUAL CONFLICT BETWEEN GENERAL RELATIVITY THEORY AND QUANTUM MECHANICS

1. We can cite the following homely example: "Think of waves on the surface of water. Here we can describe two entirely different things. Either we may observe how the undulatory surface forming the boundary between water and air alters in the course of time; or else—with the help of small floats—we can observe how the positions of separate particles of water change. If the existence of such floats for tracking the motion of the particles of a fluid . . . a fundamental impossibility in physics if, in fact, nothing else whatever were observable than the shape of the space occupied by the water as it varies with time, we should have no ground for assuming that water consists of movable particles. But all the same, we could characterize it as a medium" (A. Einstein, 1933).

CHAPTER 9. FURTHER THEORETICAL REMARKS ON THE VACUUM FLUCTUATIONS

1. It should be recalled that, as early as 1931, the physicist and astronomer Georges Lemaître had the insight to interpret the cosmological constant as originating in the behavior of matter at very high energies.

2. In 1934, Georges Lemaître used an unusual perfect-fluid equation of state to interpret the cosmological constant as being due to vacuum energy. In 1973, Edward Tryon proposed that the universe might be a large-scale quantum-mechanical vacuum fluctuation where positive mass-energy is balanced by negative gravitational potential energy. During the 1980s, there were many attempts to relate the fields that generate the vacuum energy to specific fields predicted by the grand unification theory, and to use observations of the universe to confirm that theory. These efforts have failed so far, and the exact nature of the particles or fields that generate vacuum energy, with a density such as that required by inflation theory, remains a mystery.

CHAPTER 10. MORE INTUITIVE REMARKS ON THE CASIMIR EFFECT AND FORCE, AND ON THEIR SIGNIFICANCE

1. Quantum electrodynamics (QED) is a relativistic quantum field theory of electromagnetism. QED, developed by a number of physicists in 1929, mathematically describes all phenomena involving electrically charged particles interacting by means of exchange of photons, whether the interaction is between light and matter or between two charged particles. It has been called "the jewel of physics" for its extremely accurate predictions of quantities like the anomalous magnetic moment of the electron and the Lamb shift of the energy levels of hydrogen.

In classical physics, owing to interference, light is observed to take the stationary path between two points; but how does light know where it's going? That is, if the start and end points are known, the path that will take the shortest time can be calculated. But when light is first emitted, the end point is not known, so how is it that light always takes the *quickest* path? In some interpretations, it is suggested that, according to QED, light does not have to—it simply goes over every possible path, and the observer (at a particular location) simply detects the mathematical result of all wave functions added up (as a sum of all line integrals). For other interpretations, paths are viewed as nonphysical mathematical constructs that are equivalent to other, possibly infinite, sets of mathematical expansion. According to QED, light can go slower or faster than c, but will travel at speed c on average.

Physically, QED describes charged particles (and their antiparticles) interacting with each other by the exchange of photons. The magnitude of these interactions can be computed using perturbation theory; these rather complex formulas have a remarkable pictorial representation as Feynman diagrams. The application of quantum mechanics to fields rather than single particles, resulting in what are known as quantum field theories, began in 1927. Early contributors included Dirac, W. Pauli, Weisskopf, and Jordan. This line of research culminated in the 1940s in the quantum electrodynamics of R. Feynman, F. Dyson, J. Schwinger, and S.-I. Tomonaga, which, as a theory of electrons, positrons, and the electromagnetic field, was the first satisfactory quantum description of a physical field and the creation and annihilation of quantum particles.

2. By convention, the original form of quantum mechanics (Schrödinger equation, Heisenberg matrix, . . .) is denoted "first quantization," whereas quantum field theory is formulated (by Dirac, Tomonaga, Feynman, Schwinger, Dyson, and others) in the language of "second quantization" (uncertain relations, noncommutative rules, . . .).

3. Heisenberg's "uncertainty relation" has had profound implications for such fundamental notions as causality and the determination of the future behavior of a subatomic particle, as well as philosophical implications for the conception of physical reality.

CHAPTER 11. DYNAMICAL PRINCIPLES OF INVARIANCE
AND THE PHYSICAL INTERACTIONS

1. The "Standard Model" of particle physics is a theory of three of the four known fundamental interactions, and the elementary particles that take part in these interactions. These particles make up all visible matter in the universe. Every high-energy physics experiment carried out since the mid-twentieth century has eventually yielded findings consistent with the Standard Model. Still, the Standard Model falls short of being a complete theory of fundamental interactions, because it does not include gravitation, dark matter, or dark energy. It is not quite a complete description of leptons either, because it does not describe nonzero neutrino masses, although extensions do. The Standard Model is a gauge theory of the strong ($SU(3)$) and electroweak ($SU(2) \times U(1)$) interactions with the gauge group (sometimes called the *Standard Model symmetry group*) $SU(3) \times SU(2) \times U(1)$. Again, it does not take gravitation into account.

2. In particle physics, the *electroweak interaction* is the unified description of two of the fundamental interactions of nature: electromagnetism and the weak interaction.

Although these two forces appear quite different at everyday low energies, the theory models them as two different aspects of the same force. Above the unification energy, on the order of 100 GeV, they would merge into a single electroweak force. Mathematically, the unification is accomplished under an $SU(2) \times U(1)$ gauge group. The corresponding gauge bosons are the photons of electromagnetism and the W and Z bosons of the weak force. Quantum electrodynamics (QED) is described by a $U(1)$ group, and is replaced in the unified electroweak theory by a $U(1)$ group representing a weak hypercharge rather than an electric charge. The massless bosons from the $SU(2) \times U(1)$ theory mix after spontaneous symmetry breaking to produce the three massive weak bosons, and the photon field.

3. *Quantum chromodynamics* (QCD) is a theory of the strong interaction (color force), a fundamental force describing the interactions of the quarks and gluons making up hadrons (such as the proton, neutron, or pion). It is the study of the $SU(3)$ Yang-Mills theory of color-charged fermions (the quarks). QCD is a quantum field theory of a special kind called a non-Abelian gauge theory. It is an important part of the Standard Model of particle physics.

4. Isospin is a term introduced to describe groups of particles that have nearly the same mass, such as the proton and neutron. This doublet of particles is said to have isospin 1/2, with projection +1/2 for the proton and –1/2 for the neutron. The three pions compose a triplet, suggesting isospin 1. Isospin is associated with the fact that the strong interaction is independent of electric charge. Any two numbers of the proton-neutron isospin doublet experience the same strong interaction: proton-proton, proton-neutron, neutron-neutron have the same strong force attraction. At the quark level, the up and down quarks form an isospin doublet ($I = 1/2$), and the projection +1/2 is assigned to the up quark and the projection –1/2 to the down. The strange quark, in a class by itself, has isospin $I = 0$.

CHAPTER 12. QUANTUM ELECTRODYNAMICS AND GAUGE THEORY

1. Weak isospin is the gauge symmetry of the weak interaction that connects quark and lepton doublets of lefthanded particles in all generations; for example, up and down quarks, top and bottom quarks, electrons and electron neutrinos. By contrast, (strong) isospin connects only up and down quarks, acts on both chiralities (left and right), and is a global (not a gauge) symmetry. Starting in the 1950s, attempts were made to promote isospin from a global to a local symmetry. In 1954, Chen Ning Yang and Robert Mills suggested that the notion of protons and neutrons, which are continuously rotated into each other by isospin, should be allowed to vary from point to point. To describe this, the proton and neutron direction in isospin space must be defined at every point, giving local basis for isospin. A gauge connection would then describe how to transform isospin along a path between two points. This Yang-Mills theory describes vector bosons, like the photon of electromagnetism. Unlike the photon, the $SU(2)$ gauge theory would contain self-interacting gauge bosons. The condition of gauge invariance suggests that they have zero mass, just as in electromagnetism.

2. In particle physics, a *hadron* is a composite particle (made of quarks) held together by the strong force. Hadrons are categorized in two families: baryons (made of three

quarks) and mesons (made of one quark and one antiquark). The best-known hadrons are protons and neutrons (both baryons), which can be found in the atomic nuclei. All hadrons except protons are unstable, and all undergo particle decay—although neutrons are stable when found inside the atomic nuclei.

3. Some of the symmetries referred to here are the "flavor" symmetries of (strong) isospin and its generalization to Murray Gell-Mann's unitary symmetry $SU(3)$ dating from about 1961. More precisely, Gell-Mann formulated the quark model of hadronic resonances, and identified the flavor symmetry of the light quarks, extending isospin to include strangeness, which he also discovered. In particle physics, *flavor* is a quantum number of elementary particles. In quantum chromodynamics, flavor is a global symmetry. In the electroweak theory, by contrast, this symmetry is broken, and the flavor-changing process persists, as for example quark decay or neutrino oscillations. Quantum chromodynamics contains six flavors of quarks, though their masses differ. As a result, they are strictly interchangeable with each other. The up and down flavors are close to having equal masses, and the theory of these two quarks calls for their possessing an approximate $SU(2)$ symmetry (isospin symmetry). Under some circumstances, one can take N_f flavors to have the same masses, and thus obtain an effective $SU(N_f)$ flavor symmetry.

CHAPTER 13. VACUUM AS THE SOURCE OF ASYMMETRY

1. Goldstone's theorem states that whenever a continuous symmetry is spontaneously broken, new massless scalar particles (or light, if the symmetry was not exact) appear in the spectrum of possible excitations (of the vacuum). There is one scalar particle—called a Goldstone boson—for each generator of the symmetry that is broken, i.e., that does not preserve the ground state. The theorem can also be formulated in another manner. It states that there exist nonvacuum states with arbitrarily small energies. Take, for example, a chiral $N = 1$ super QCD model with a nonzero squark VEV that is conformal in the IR. The chiral symmetry is a global symmetry that is (partially) broken spontaneously. Some of the "Goldstone bosons" associated with this spontaneous symmetry breaking are charged under the unbroken gauge group and, hence, the composite bosons have a continuous mass spectrum with arbitrarily small masses, yet there is no Goldstone boson with exactly zero mass. In other words, the Goldstone bosons are infraparticles. In theories with gauge symmetry, the Goldstone bosons are "eaten" by the gauge bosons. In the process, the latter become massive and their new longitudinal polarization is provided by that of the Goldstone bosons.

2. Recall that a *Hermitian form on a left R-module X* is a mapping $\phi: X \times X \to R$ that is linear in the first argument and satisfies the condition

$$\phi(y, x) = \phi(y, x)^J, \quad x, y \in X.$$

Here is a ring with a unit element, a ring equipped with an involutory anti-automorphism J. In particular, ϕ is a sesquilinear form on X. The module X itself is then called a *Hermitian space*. By analogy with what is done for bilinear forms, equivalence is defined for Hermitian forms and, correspondingly, isomorphism of Hermitian spaces (in particular, automorphism). All automorphisms of a Hermitian form ϕ form a group $U(\phi)$, which is called the unitary group associated with the Hermitian form ϕ; its structure has been well studied where R is a skew-field. A Hermitian form is a special case of an ε-Hermitian form (where ε is an element in the center of R), that is, a sesquilinear form ψ on X for which

$$\psi(x, y) = \varepsilon\, \psi(y, x)^J, \ x, y \in X.$$

When $\varepsilon = 1$, and ε-Hermitian form is Hermitian, and when $\varepsilon = -1$, the form is called skew-Hermitian or anti-Hermitian. If $J = 1$, a Hermitian form is a symmetric bilinear form, and a skew-Hermitian form is a skew-symmetric or antisymmetric bilinear form. If the mapping

$$X \to \mathrm{Hom}_R(X, R), \qquad y \mapsto f_y,$$

where $f_y(x) = \phi(x, y)$ for any $x \in X$, is bijective, then ϕ is called a nondegenerate Hermitian form or a Hermitian scalar product on X. If X is a free R-module with a basis $e_1, \ldots, e_n$, then the matrix $\| a_{ij} \|$, where $a_{ij} = \phi(e_i, e_j)$, is called the matrix of R in the given basis; it is a Hermitian matrix (that is, $a_{ij} = a_{ij}^J$). A Hermitian form ϕ is nondegenerate if and only if $\| a_{ij} \|$ is invertible. If R is a skew-field, if $cha\,R \neq 2$, and if X is finite-dimensional over R, then X has an orthogonal basis relative to ϕ (in which the matrix is diagonal). If R is a commutative ring with identity, if $R_0 = \{ r \in R : r^J = r \}$, and if the matrix of ϕ is definite, then its determinant lies in R_0. Under a change of basis in X, this determinant is multiplied by a nonzero element of R of the form $\alpha\, \alpha^J$, where α is an invertible element of R. The determinant defined up to multiplication by such elements is called the determinant of the Hermitian form, or of the Hermitian space X; it is an important invariant, and is used in the classification of Hermitian forms. Let R be commutative. Then a Hermitian form ϕ on X gives rise to a quadratic form $Q(x) = \phi(x, x)$ on X over R_0. The analysis of such forms lies at the basis of the construction of the Witt group of R with an involution. When R is a maximal ordered field, then the law of inertia extends to Hermitian forms (and there arise the corresponding concepts of the signature, the index of inertia, and positive and negative definiteness). If R is a field and $J \neq 1$, then R is a quadratic Galois extension of R_0, and isometry of two nondegenerate Hermitian forms over R is equivalent to isometry of the quadratic forms over R_0 generated by the Hermitian forms; this reduces the classification of nondegenerate Hermitian forms over R to that of nondegenerate quadratic forms over R_0. If $R = C$ and J is the involution of complex conjugation; then a complete system of invariants of Hermitian forms over a finite-dimensional space is given by the rank and the signature of the corresponding quadratic forms. If R is a local field or the field of functions of a single variable over a finite field, then a complete system of invariants for nondegenerate Hermitian forms is given by the rank and the determinant. If R is a finite field, then there is only one invariant, the rank. There is also the case where R is an algebraic extension of Q. Charles Hermite was the first, in 1853, to consider the forms that bear his name in connection with certain problems of number theory.

CHAPTER 16. CREATION OF UNIVERSES FROM NOTHING

1. According to S. Hawking, quantum theory causes black holes to radiate and lose mass. The *no-hair theorem* shows that a great amount of information is lost when a body (e.g., a star) collapses to form a black hole. This loss of information would introduce a new level of uncertainty into physics, over and above the usual uncertainty associated with Heisenberg's uncertainty principle.

CHAPTER 17. STRING LANDSCAPE AND VACUUM ENERGY

1. Recall that superstring theory inevitably introduces *branes*. D-branes are hypersurfaces on which superstrings can end. (Here, "D" stands for Dirichlet boundary condition, and "brane" for membrane.) Superstrings can be open or closed. Open superstrings (photons) can propagate only on the $p + 1$-dimensional brane, whereas closed superstrings (gravitons) can be lacking or far away from the brane. In both cases, we are living in the branes. A p-brane is a space-time object that is a solution of the Einstein equation in the low-energy limit of superstring theory, with the energy density of the nongravitational fields confined to some p-dimensional subspace of the nine space dimensions in the theory. (Remember, superstring theory lives in ten space-time dimensions, which means one time dimension plus nine space dimensions.) For example, in a solution with electric charge, if the energy density in the electromagnetic field is distributed along a line in space-time, this one-dimensional line would be considered a p-brane with $p = 1$. D-branes are important in understanding black holes in string theory, especially in counting the quantum states that lead to black hole entropy, which has been a huge accomplishment for string theory.

2. It may be historically useful to emphasize that Einstein introduced the cosmological constant to his (field) equations of general relativity because he believed the universe to be static. Faced with evidence that the universe was actually expanding, however, he decided to remove it, later referring to the cosmological constant as the biggest blunder of his life. Recent observations suggest that the expansion of the universe is in fact accelerating, which favors a small, but nonzero, positive cosmological constant with a value of 10^{-120} in Planck units. But perhaps Einstein's most serious mistake regarding the cosmological constant, Λ, was actually to believe that he had the right to decide whether or not it should be included in his equations in the first place. Even if Λ is very small or zero, its value must be explained, and the explanation will very likely entail the vacuum's energy and fluctuations issue as a key feature. This is one of the greatest puzzles of modern theoretical physics. Without a cosmological constant, the most symmetric solution of Einstein's equations in the vacuum is the flat, four-dimensional Minkowskian space-time of special relativity. In 1917, the Dutch astronomer Willem de Sitter produced the analogous solution if the cosmological constant is in fact nonzero. If Λ is positive, the solution is called "de Sitter space" and Λ is the vacuum energy that curves space-time; a negative value of Λ corresponds to what is called anti-de Sitter space. (Anti-de Sitter space is a maximally symmetric Lorentzian manifold with constant negative scalar curvature; it is the Lorentzian analogue of n-dimensional hyperbolic space, just as Minkowskian space

and de Sitter space are the analogues of Euclidean and elliptical spaces, respectively.) Einstein immediately rejected the de Sitter solution, because it went against his intuition—it implied that space-time can be curved in the absence of matter—although in the end, after some debate, he accepted the idea. In de Sitter space, the universe expands exponentially, which is the basis of the inflationary model of the universe. De Sitter space also plays an important role in string theory.

CHAPTER 18. CONCLUDING REMARKS

1. In some sense, one can say that chaos theory, a mathematical subdiscipline that studies complex systems, could not have emerged before the discovery of quantum mechanics and the consequent criticism of the deterministic paradigm. Developed in the 1960s by E. Lorentz and other physicists, chaos theory allows us to demonstrate that some things are not linearly caused by other things, and that, therefore, future events cannot be reliably predicted from past events. Many natural phenomena, such as the global weather system, will not let themselves be predicted. According to chaos theory, the unfolding of such natural phenomena is inherently uncertain, because of their sensitivity to initial conditions; small changes at a particular moment may lead to large differences later. Thus, the initial situation of a complex nonlinear system cannot be exactly determined, and the evolution of such a system, therefore, cannot be accurately predicted.

The behavior of a chaotic system over long time scales is: (1) unpredictable (i.e., one cannot forecast in detail and with precision the evolution of the system), (2) seemingly random but not arbitrarily so, since it can exhibit ordered patterns, (3) sensitive to dependence on initial conditions (minor changes can cause huge fluctuations and magnified effects in the behavior of the system), (4) vulnerable to the presence of bifurcations, which are sudden changes in a system that occur when it is slightly modified, and (5) characterized by a *strange attractor* that is often a *fractal*. Chaotic behavior differs from random behavior in the sense that it is not completely random, and the strange attractor governs its structure. To sum up, chaos theory is the qualitative study of unstable aperiodic behavior in nonlinear dynamical systems; the nonlinearity of the equations usually renders a closed-form solution impossible. Researchers into chaotic phenomena thus seek a qualitative account of the behavior of nonlinear differentiable dynamical systems. After modeling a physical system with a set of equations, they do not concentrate on finding a formula that will make possible the exact prediction of a future state from a present or past one. Instead, they use mathematical methods to provide some idea about the long-term behavior of the solutions.

For every set S in a Euclidean space, there exists a real value D such that the d-dimensional Hausdorff measure is infinite for $d < D$ and vanishes for $d > D$. This D is called the Hausdorff dimension. A *fractal* is defined as a set for which the Hausdorff dimension strictly exceeds the dimension of the set, which is topologically defined in a Euclidean space. There are many interesting examples of fractals having fractional dimensions.

Let A be a non-empty closed invariant set. A is called an *attractor* if it has an open neighborhood U satisfying the following conditions: (i) U is positively invariant; (ii) For each open neighborhood V of A, there is a $T > 0$ such that $ft(U) \in V$ for all $t \in T$. Condition

(ii) implies that $ft > 0 \in t(U) = A$ and $f(x) \in A$ for all $x \in U$. Thus an attractor is asymptotically stable. If A is an attractor, the basin of A is the union of all open neighborhoods of A satisfying (i) and (ii).

In more intuitive terms, an attractor is a set toward which a dynamical system evolves over time. That is, points that get close enough to the attractor remain close even if slightly disturbed. Geometrically, an attractor can be a point, a curve, a manifold, or even a complicated set with a fractal structure known as *strange attractor*. An attractor is described as strange if it has non-integer dimension. This is often the case when the dynamics on it are chaotic, but there exist also strange attractors that are not chaotic. The term "strange" was coined by David Ruelle and Floris Takens to describe the attractor that resulted from a series of bifurcations of a system describing fluid flow. Strange attractors are often differentiable in a few directions, but some are like a Cantor dust, and therefore not differentiable. Strange attractors are unique from other phase-space attractors in the sense that one does not know exactly where on the attractor the system will be. Two points on the attractor that are near each other at one time will be arbitrarily far apart at later times. The only restriction is that the state of the system remains on the attractor. Strange attractors are also unique in that they never close in on themselves—the motion of the system never repeats (it is nonperiodic).

Bibliography

Abel, S. A., J. Jaeckel, and V. V. Khoze, 2007. "Why the early Universe preferred the non-supersymmetric vacuum: Part II," *Journal of High Energy Physics*; hep-th/0611130.

Adler, S. L., 1980. "Ordered-R vacuum action functional in scalar-free unified theories with spontaneous scale breaking," *Phys. Rev. Lett.*, 44, 1567–82.

——, 1994. "Generalized quantum dynamics," *Nucl. Phys. B*, 415, 195–226.

Aguilar, M. A., J. M. Isidro, and M. Socolovsky, 2004. "Moduli space of potentials in the Aharonov-Bohm effect," *Advances in Applied Clifford Algebras*, 14:179–84.

Aharonov, Y., 1986. "Non-local phenomena and the Aharonov-Bohm effect," pp. 41–56 in *Quantum Concepts in Space and Time*, R. Penrose and C. J. Isham (eds.). Clarendon Press, Oxford.

Aharonov, Y. and D. Bohm, 1959. "Significance of electromagnetism potentials in the quantum theory," *Phys. Rev.*, 115(3), 485–91.

Aitchison, I. J. R., 1985. "'Nothing's plenty': The vacuum in modern quantum field theory," *Contemp. Phys.*, 26(4), 333–91.

Albeverio, S., J. Jost, S. Paycha, and S. Scarlatti, 1997. *A Mathematical Introduction to String Theory, Variational Problems, Geometric and Probabilistic Methods.* Cambridge University Press, Cambridge.

Albrecht, A., and P. J. Steinhardt, 1982. "Cosmology for grand unified theories with radioactively induced symmetry breaking," *Phys. Rev. Lett.*, 48, 1220–23.

Alexander, H. G. (ed.), 1956. *The Leibniz-Clarke Correspondence.* Manchester University Press, Manchester.

Allen, B., 1985. "Quantum states in de Sitter space," *Phys. Rev. D*, 32(12), 3136–52.

Allen, R. E., Oct. 2001. "Dark matter, quantum gravity, vacuum energy, and Lorentz invariance," arXiv:hep–th/0110208.

Ambjorn, J., and S. Wolfraum, 1983. "Properties of the vacuum. Part 1, mechanical and thermodynamic," *Ann. Phys.*, 147, 1–32; Part. 2. Electrodynamics, 33–56.

Anderson, A., and B. DeWitt, 1988. "Does the topology of space fluctuate?" pp. 74–88 in *Between Quantum and Cosmos*. Studies and Essays in Honor of John Archibald Wheeler; W. H. Zurek, A. van der Merwe, and W. A. Miller (eds.). Princeton University Press, Princeton.

Andrews, D. L. et al., 2001. "Conceptualization of the Casimir effect," *Eur. J. Phys* 22, 447–51.

Arnold, V., 1978. *Mathematical Methods of Classical Mechanics.* Springer-Verlag, New York.

Atiyah, M. F., 1979. *Geometry of Yang-Mills Fields*, Accademia Nazionale dei Lincei. Edizioni della Scuola Normale Superiore, Pise.

———, 1988. "New invariants of three and four dimensional manifolds," in *The Mathematical Heritage of Hermann Weyl*, Proc. Symp. Pure Math., 48, American Math. Soc., 285–329.

Atkatz, D., and H. Pagels, 1982. "Origin of the universe as a quantum tunneling event," *Phys. Rev. D*, 25, 2065–73.

Balian, R., 2003. "Casimir effect and geometry," in *Poincaré Seminar 2002: Vacuum Energy— Renormalization*, B. Duplantier and V. Rivasseau (eds.), 71–92. Birkhäuser, Basel.

Barcelos-Neto, J., and E. C. Marino, 2002. "A new sequence of topological terms at any space-time dimensions," *Europhysics Letters*, 57(4), 473–98.

Barret, T. W., 1989. "On the distinction between fields and metric," *Annales de la Fondation Louis de Broglie*, 14(1), 37–75.

Barrow, J. D., 2002. *The Book of Nothing: Vacuums, Voids, and the Latest Ideas about the Origins of the Universe*. Vintage, London.

Barrow, J. D., and J. Silk, 1983. *The Left Hand of Creation*. Basic Books, New York.

Beck, C., 2002 *Spatio-Temporal Chaos and Vacuum Fluctuations of Quantized Fields*, Advanced Series in Nonlinear Dynamics, vol. 21. World Scientific, Singapore.

Belavin, A. A., and A. M. Polyakov, 1977. "Quantum fluctuations of pseudoparticles," *Nucl. Phys. B*, 123(3), 429–44.

Belavin, A., A. Polyakov, A. Schwartz, and Y. Tyupkin, 1975. "Pseudo-particle solutions of the Yang-Mills equations," *Phys. Lett. B*, 59, 85–87.

Bell, J. S., 1987. *Speakable and Unspeakable in Quantum Mechanics*. Cambridge University Press, Cambridge.

Berry, M. V., 1987. "Quantum chaology," pp. 183–98 in M. V. Berry, I. C. Percival, and N. O. Weiss, eds., *Dynamical Chaos*. Princeton University Press, Princeton.

Binetruy, P., E. A. Dudas, and F. Pillon, 1994. "The vacuum structure in a supersymmetric gauged Nambu-Jona-Lasinio model," *Nucl. Phys. B*, 415, 175–94.

Birrel, N. D., and P. C. W. Davies, 1982. *Quantum Fields in Curved Space*. Cambridge University Press, Cambridge.

Blohincev, D. I., 1974. "Geometry and physics of the microcosmos," *Sov. Phys. Usp.*, 16, 485–93.

Bohm, D., 1951. *Quantum Theory*. Prentice-Hall, New York.

Bohm, D., and B. J. Hiley, 1993. *The Undivided Universe: An Ontological Interpretation of Quantum Theory*. Routledge, London.

Bohr, N., 1913. "On the constitution of atoms and molecular, Part I," *Philosophical Magazine*, 26, 1–24.

Boi, L., 1992. "L'espace: concept abstrait et/ou physique; la géométrie entre formalisation mathématique et tude de la nature," pp. 65–90 in *1830–1930: A Century of Geometry, Epistemology, History and Mathematics*, L. Boi, D. Flament, and J.-M. Salanskis (eds.), Lecture Notes in Physics 402, Springer-Verlag, Berlin.

———. 2000. "Géométrie de l'espace-temps et nature de la physique: Quelques réflexions historiques et épistémologiques," *Manuscrito*, 23 (1), 31–97.

————, 2004a. "Geometrical and topological foundations of theoretical physics: From gauge theories to string program," *Internat. J. Math. Sci.*, 34, 1777–1836.

————, 2004b. "Theories of space-time in modern physics," *Synthese*, 139, 429–89.

————, 2005. "Topological knots models in physics and biology," in: *Geometries of Nature, Living Systems and Human Cognition. New Interactions of Mathematics with the Natural Sciences and Humanities*, L. Boi (ed.), (203–78). World Scientific, Singapore.

————, 2006a. "From Riemannian geometry to Einstein's general relativity theory and beyond: Space-time structures, geometrization and unification," pp. 1066–75 in *Proceedings of the Albert Einstein Century International Conference*, J.-M. Alimi and A. Füzfa (eds.). American Institute of Physics. Melville, N.Y.

————, 2006b. "Mathematical knot theory," in: *Encyclopedia of Mathematical Physics*, J.-P. Françoise, G. Naber, and T. S. Tsun (eds.), (399–407). Elsevier, Oxford.

————, 2006c. "Nouvelles dimensions mathématiques et épistémologiques du concept d'espace en physique relativiste et quantique," in *L'espace physique entre mathématique et philosophie*, 101–33, M. Lachièze-Rey (ed.). EDP Sciences, Paris.

————, 2006d. "The aleph of space. On some extension of geometrical and topological concepts in twentieth-century mathematics: From surfaces and manifolds to knots and links," pp. 79–152 in *What Is Geometry?*, G. Sica (ed.), Polimetrica, Milan.

————, 2009a. "Clifford geometric algebras, spin manifolds, and group action in mathematical physics," *Advances in Applied Clifford Algebras* 19(3–4), 611–56.

————, 2009b. "Geometria e dinamica dello spazio-tempo nelle teorie fisiche recenti. Su alcuni problemi coincettuali della fisica contemporanea," *Giornale di Fisica*, Società Italiana di Fisica, 5 (suppl. 1), 1–10.

————, 2009c. "The geometric foundations of contemporary physics and reflections on the nature of space-time," *Foundations of Physics* (to appear in 2012).

————, 2009d. "The geometric foundations of contemporary physics, and some reflections about the nature of space-time," pp. 1–36 in *Proc. XVIII International Symposium SIGRAV*, Italian Society of General Relativity, M. Francaviglia (ed.).

————, 2009e. "Ideas of geometrization, geometric invariants of low-dimensional manifolds, and topological quantum field theories," *International Journal of Geometric Methods in Modern Physics*, 6(5), 1–57.

————, 2011a (forthcoming). "Some reflections on the relationship between geometry and reality, space-time theory and the geometrization of theoretical physics, from B. Riemann to H. Weyl and beyond," *Foundations of Science*.

————, 2011b. (forthcoming). "Topological structures in classical and quantum physics," *Journal of Topology and Its Applications*.

————, 2011c. "Non-commutative geometry and the structures of space-time," *Journal of Advances in Mathematical Physics*, special issue on "Nonlinear and noncommutative mathematics: New developments and applications in quantum physics."

Bott, R., and C. Taubes, 1994. "On the self-linking number of knots," *J. Math. Phys.*, 35(10), 5247–87.

Bott, R., and L. W. Tu, 1982. *Differential Forms in Algebraic Topology*. Springer-Verlag, New York.

Boyer, T. H., 1968. "Quantum electromagnetic zero-point energy of a conducting spherical shell and the Casimir model for a charged particle," *Phys. Rev.*, 174(5), 1764–76.

———, 1989. "Conformal symmetry of classical electromagnetic zero-point radiation,"
Found. Phys., 19(4), 349–65.

Brading, K. A., and E. Castellani (eds.), 2003. *Symmetries in Physics: Philosophical Reflections.* Cambridge University Press, Cambridge.

Brandenberger, R., 2001. "On the spectrum of fluctuations in an effective field theory of
ekpyrotic universe," *Journal of High Energy Physics*, 11, 1088–1126.

Bros, J., H. Epstein, and U. Moschella, 2002. "Towards a general theory of quantized
fields on the anti-de Sitter space-time," *Commun. Math. Phys.*, 231, 481–528.

Brout, R., F. Englert, and E. Gunzig, 1978. "The creation of the universe as a quantum
phenomenon," *Annals of Physics*, 115: 78–106.

Brouwer, L. E. J. 1912. "Über Affildanger non Mannigfaltigkeiten, *Math. Ann.*, 71,
379–94.

Brown, H. R., and R. Harré (eds.), 1988. *Philosophical Foundations of Quantum Field Theory.* Clarendon Press, Oxford.

Cabibbo, N., 1963. "Unitary symmetry and leptonic decays," *Phys. Rev. Lett.*, 10, 531–33.

Callan, C., and S. Coleman, 1977. "Fate of the false vacuum. II, First quantum corrections," *Phys. Rev. D*, 16, 1762–68.

Callender, C., and N. Hugget (eds.), 2001. *Physics Meets Philosophy at the Planck Scale.*
Cambridge University Press, Cambridge.

Candelas, P., G. T. Horowitz, A. Strominger, and E. Witten, 1985. "Vacuum configurations for superstrings," *Nucl. Phys. B*, 258, 46–74.

Cao, T. Yu., 1997. *Conceptual Developments of 20th Century Physics.* Cambridge University
Press, Cambridge.

Casimir, H. B. G., 1948. "On the attraction between two perfectly conducting plates,"
Proc. Koninkl. Ned. Akad. Wetenschappen B, 51(7), 793–96.

Casimir, H. B. G., and D. Polder, 1948. "The influence of retardation on the London-van
der Waals forces," *Phys. Rev.*, 73(4), 360–72.

Cassé, M., 1993. *Du vide et de la creation.* Odile Jacob, Paris.

Cheng, T.-P., and L.-F. Li, 1984. *Gauge Theory of Elementary Particle Physics.* Clarendon
Press, Oxford.

Chern, S. S., 1989. "Vector bundles with a connection," *Global Differential Geometry*, S. S.
Chern (ed.), MAA Studies in Mathematics, vol. 27, Mathematical Association of America, 1–26.

Chern, S.-S., and J. Simons, 1974. "Characteristics forms and geometric invariants," *Annals of Mathematics*, 99(1), 48–69.

Cho, Y. M., 2002. "Knot topology of QCD vacuum," *Phys. Lett. B*, 644, 208–11.

Cho, Y. M., and P. G. O. Freund, 1975a. "Gravitating 't Hooft monopoles," *Phys. Rev. D*,
12, 1588–89.

———, 1975b. "Non-Abelian gauge fields as Nambu-Goldstone fields," *Phys. Rev. D*, 12,
1711–20.

Clifford, W. K., 1982. "On the space-theory of matter," *Mathematical Papers*, R. Tucker
(ed.), London. Macmillan, New York.

Coleman, S., 1966. "The invariance of the vacuum is the invariance of the world," *J. Math. Phys.*, 7, 787–812.

———, 1975. "Secret symmetry: An introduction to spontaneous symmetry breakdown and gauge fields," pp. 138–215 in *Laws of Hadronic Matter*, A. Zichichi (ed.). Academic Press, New York.

———, 1985. *Aspects of Symmetry*, Selected Erice lectures. Cambridge University Press, Cambridge.

———, 1988. "Why there is nothing rather than something: A theory of the cosmological constant," *Nucl. Phys. B*, 310, 643–58.

Connes, A., and A. H. Chamseddine, 2006. "Inner fluctuations of the spectral action," *J. Geom. Phys.*, 57, 1–21.

Cremmer, E., and J. Scherk, 1977. "Spontaneous compactification of extra space dimensions," *Nucl. Phys. B*, 118(1–2), 61–75.

Crowell, L. B., 2005. *Quantum Fluctuations of Spacetime*. World Scientific, Singapore.

Cushing, J. T., 1998. *Philosophical Concepts in Physics*. Cambridge University Press, Cambridge.

Dancoff, S. M., 1939. "On radiative corrections for electron scattering," *Phys. Rev.*, 55, 959–63.

Daniel, C. C., and H. E. Puthoff, 1993. "Extracting energy and heat from the vacuum," *Phys. Rev. E*, 48(2), 1562–65.

Davies, P. (ed.), 1989. *The New Physics*. Cambridge University Press, Cambridge.

Davies, P. C. W., 2005. "Quantum vacuum friction," *J. Opt. B: Quantum Semiclass. Opt.* 7, 1464–66.

De Broglie, L., 1937. *La physique nouvelle et les quanta*. Flammarion, Paris.

Delamotte, B., 2004. "A hint of renormalization," *Amer. J. Phys.*, 72, 170–84.

Descartes, R., 1644. Les principes de la philosophie, in *Œuvres Philosophiques*, F. Alquié (ed.), 3 vols. G.-F. Flammarion, Paris, 1963–73.

De Sitter, W., 1917. "On the curvature of space," *Proc. Kon. Ned. Acad. Wet.*, 20, 229–43.

DeWitt, B. S., 1975. "Quantum field theory in curved spacetime," *Phys. Rep.*, 19, 295–357.

———, 1979. "Quantum gravity: The new synthesis," pp. 680–745 in *General Relativity. An Einstein Centenary Survey*, S.W. Hawking and W. Israel (eds.). Cambridge University Press, Cambridge.

———, 2003. *The Global Approach to Quantum Field Theory*. Clarendon Press, of Oxford University Press, Oxford.

Dirac, P. A. M., 1928. "The quantum theory of the electron," *Proc. Roy. Soc.*, 117, 610–24.

———, 1930a. "A theory of electrons and protons," *Proc. Roy. Soc. A*, 126, 360–65.

———, 1930b. *The Principles of Quantum Mechanics*. Clarendon Press, Oxford.

———, 1931. "Quantized singularities in the electromagnetic field," *Proc. Roy. Soc. A*, 133, 60–72.

———, 1933. "Théorie du positron," pp. 203–12 in *Rapport du 7ᵉ Conseil Solvay de Physique, Structure et Propriétés des Noyaux Atomiques*. Gauthier-Villars, Paris.

Dolgov, A. D., and Ya. B. Zeldovich, 1981. "Cosmology and elementary particles," *Rev. Mod. Phys.*, 53, 1–41.

Donaldson, S. K., 1983. "An application of gauge theory to four-dimensional topology," *J. Diff. Geometry*, 18, 279–315.

Donaldson, S. K., and P. B. Kronheimer, 1990. *The Geometry of Four-Manifolds*. Oxford University Press, Oxford.

Dowker, F., J. P. Gauntlett, S. B. Giddings, and G. T. Horowitz, 1994. "On pair creation of extremal black holes and Kaluza-Klein monopoles," *Phys. Rev.* D, 50(4), 2662–79.

Duplantier, B., and V. Rivasseau (eds.), 2003. *Poincaré Seminar 2002: Vacuum Energy, Renormalization*, Progress in Mathematical Physics Series, Vol. 30. Birkhäuser, Basel/Boston.

Dyson, F. J., 1949a. "The radiation theories of Tomonaga, Schwinger and Feynman," *Phys. Rev.*, 75, 486–502.

———, 1949b. "The *S*-matrix in quantum electrodynamics," *Phys. Rev.*, 75, 1736–55.

Einstein, A., 1905. "Zur Electrodynamik bewegter Körper," *Annalen der Physik*, 17, 891–921.

———, 1922. *Ether and the Theory of Relativity*. Methuen and Co., London.

———, 1923. *Fundamental Ideas and Problems of the Theory of Relativity*. Elsevier, Amsterdam.

———, 1933. *Mein Weltfild*. Querido Verlag, Amsterdam.

Enss, V., and R. Weder, 1995. "The geometrical approach to multidimensional inverse scattering," *J. Math. Phys.*, 36, 3902–21.

Faddeev, L., A. and J. Niemi, 1997. "Knots and particles," *Nature*, 58, 387–96.

Fahr, H. J., 1998. "The modern concept of vacuum and its relevance for the cosmological models of the universe," in *Philosophy of the Natural Sciences*, P. R. Weingarten and G. Schurz (eds.). Holder-Pichler-Tempspy, Vienna.

Fahr, H. J., and M. Heyl, 2007. "Cosmic vacuum energy decay and creation of cosmic matter," *Naturwissenschaften*, 94(9), 709–24.

Fakir, R., 1990. "Quantum creation of universes with non-minimal couplings," *Phys. Rev.* D, 41, 3012–23.

Faraday, M., 1839–1855. *Experimental Researches in Electricity*. R. and J. E. Taylor, London.

Felsager, B., 1981. *Geometry, Particles and Fields*. Odense University Press, Odense.

Feynman, R., 1949. "Space-time approach to quantum electrodynamics," *Phys. Rev.*, 76, 769–89.

———, 1967. *The Character of Physical Laws*. The MIT Press, Cambridge, Mass.

———, 1985. *QED: The Strange Theory of Light and Matter*. Princeton University Press, Princeton.

Fock, V. A., 1932. "Konfigurationsraum und Zweite Quantelung," *Zeit. Phys.*, 75, 622–47.

Frankel, T., 1997. *The Geometry of Physics. An Introduction*. Cambridge University Press, Cambridge.

Friedman, J. L., 1998. "Lorentzian universes from nothing," *Class. Quantum Grav.*, 15, 2639–44.

Friedmann, A., 1922. "Über die Krümmung des Raumes," *Zeitschrift für Physik*, 10(1), 377–86.

Fröhlich, J., 1992. "Spontaneously broken and dynamically enhanced global and local symmetries," *Non-Perturbative Quantum Field Theory (Selected Papers of Jürh Fröhlich)*, (193–212). World Scientific, London.

Gasperini, M., and G. Veneziano, 2003. "The pre-big bang scenario in string cosmology," *Phys. Reports*, 373, 1–212.

Gellman, M., 1962. "Symmetries of baryons and mesons," *Phys. Rev.* 125, 1076–84.

Gellman, M., and M. L. Goldberger, 1954. "The scattering of low-energy photons by particles of spin 1/2," *Phys. Rev.* 96, 1433–38.

Genz, H., 1988. *Die Entdeckung des Nichts. Leere und Fülle im Universum.* Hanser, München. (English edition: *Nothingness: The Science of Empty Space.* Basic Books, New York).

Ghilencea, D. M., D. Hoover, C. P. Burgess, and F. Quevedo, 2004. "Casimir energies for 6D supergravities compactified on $T_2/\mathbb{Z}_N$ with Wilson lines," DAMTP-2004–67, McGill-04/23.

Gibbons, G. W., M. B. Green, and J. P. Malcolm, 1996. "Instantons and seven-branes in type IIB superstring theory," *Phys. Lett. B*, 370, 37–44.

Goldstone, J., A. Salam, and S. Weinberg, 1962. "Broken symmetries," *Phys. Rev.*, 127(3), 965–70.

Gompagno, G., R. Passante, and F. Persico, 1995. *Atom-Field Interactions and Dressed Atoms.* Cambridge University Press, Cambridge.

Gott, J. R., 1982. "Creation of open universes from de Sitter space," *Nature*, 295, 304–7.

Grandy Jr., W. T., 1991. "The explicit nonlinearity of quantum electrodynamics," pp. 149–64 in *The Electron: New Theory and Experiment*, D. Hestenes and A. Weingartshofer (eds.). Kluwer Academic Publishers, Boston.

Grishchuk, L. P., and Ya. B. Zel'dovich, 1982. "Complete cosmological theories," pp. 71–86 in *Quantum Gravity (Proceedings of the 2nd Seminar on Quantum Gravity*, Moscow, 13–15 October 1981), M. A. Kharkov and P. C. West (eds.). Nauka, Moscow/Plenum Press, New York.

Grosche, C., 1993. "An introduction into the Feynman path integral," arXiv:hep-th/9302097v1 20 Feb 1993.

Gunzig, E., and S. Diner (eds.), 1998. *Le vide. Univers du tout et du rien.* Editions Complexe, Bruxelles.

Guth, A. H., 1981. "Inflationary universe: A possible solution to the horizon and flatness problems," *Phys. Rev. D*, 23, 347–56.

———, 1997. *The Inflationary Universe: The Quest for a New Theory of Cosmic Origins.* Addison-Wesley, Reading, Mass.

Guth, A. H., and D. I. Kaiser, 2005. "Inflationary cosmology: Exploring the universe from the smallest to the largest scales," *Science*, 307(5711), 884–90.

Haisch, B., A. Rueda, and Y. Dobyns, 2001. "Inertial mass and the quantum vacuum fields," *Ann. Phys.*, 10(5), 393–414.

Hartle, J. B., and S. W. Hawking, 1983. "Wave function of the universe," *Phys. Rev. D*, 28, 2960–75.

Hawking, S. W., 1975. "Particle creation by black holes," *Commun. Math. Phys.*, 43, 199–220.

———, 1996. "Singularities in the universe," *Phys. Rev. Lett.*, 17, 444–45.

Hawking, S. W. and N. Turok, 1998. "Open inflation without false vacua," *Phys. Lett.* [hep-th/9802030].

Healey, R., 2001. "On the reality of gauge potentials," *Philosophy of Science*, 68(4), 432–55.

Heisenberg, W., 1927. "Über den anschaulichen Imbalt der quantentheoretischen Kineinatik und Mechanik," *Zeitschrift für Physik*, 43, 172–98.

——, 1930. *The Physical Principles of the Quantum Theory*. University of Chicago Press, Chicago.

——, 1934. "Bemerkungen zur Diracschen Theorie des Positrons," *Zeit. Phys.*, 90, 209–31.

——, 1979. *Philosophical Problems of Quantum Physics*. Ox Bow Press, Woodbridge, Conn.

Heisenberg, W., and W. Pauli, 1929. "Zur Quantendynamik der Wellenfelder," *Zeitschrift für Physik*, 26, 1–61.

Higgs, P. W., 1964. "Broken symmetries and the mass of gauge bosons," *Phys. Rev. Lett.*, 13, 508–9.

Hilbert, D., 1924. *Grundzüge einer allgemeinen Theorie der linearen Integralgleichungen*. Teubner, Leipzig.

Hindmarsh, M. B., and T.W. B. Kibble, 1995. "Cosmic strings," *Reports on Progress in Physics*, 58(5), 477–562.

Hirsch, M., 1985. "The chaos of dynamical systems," pp. 189–96 in P. Fischer and W. R. Schmith, eds., *Chaos, Fractals, and Dynamics*. Marcel Dekker, New York.

Hodge, W. V. D., 1952. "The topological invariants of algebraic varietles," pp. 181–92 in *Proc. International Congress of Mathematicians* (Edinburgh, 1950), Amer. Math. Soc., Providence, R.I.

Hong, J., A. Vilenkin, and S. Winitzki, 2003. "Particle creation in a tunneling universe," *Phys. Rev. D*, 68(2), 1103–24.

Hopf, H., 1931. "Über die Abbildungen der 3-Sphäre auf die Kugelflache," *Math. Ann.*, 104, 637–65.

Hurewicz, W., 1941. "On duality theorems," *Bull. Amer. Math. Soc.*, 47, 562–63.

Hut, P., and M. J. Rees, 1983. "How stable is our vacuum," *Nature*, 302, 508–9.

Iqbal, A., C. Vafa, N. Nekrasov, and A. Okounhov, 2008. "Quantum foam and topological strings," *Journal of High Energy Physics*, 11, 1–44.

Isenberg, J., R. Mazzeo, and D. Pollack, 2002. "On the topology of vacuum space-time," arXiv:gr–qc/0206034v2 10 Oct 2002.

Isham, C. J., 1984. "Topological and global aspects of quantum field theory," *Relativity, Groups and Topology II*, B. S. DeWitt and R. Stora (eds.), (pp. 1059–1290). Les Houches, North-Holland, Amsterdam.

Itzykson, C., and J.-B. Zuber, 1985. *Quantum Field Theory*. McGraw-Hill, New York.

Jaekel, M.-T., and S. Reynaud, 1995. "Quantum fluctuations of vacuum stress tensors and spacetime curvature," *Annalen der Physik*, 4, 68–86.

——, 1997. "Movement and fluctuations of the vacuum," arXiv:quant-ph/9706035v1 16 Jun 1997.

Jaffe, A., and C. Taubes, 1980. *Vortices and Monopoles*. Birkhäuser, Boston.

Jahn, O., 2000. "Instantons and monopoles in general Abelian gauges," *J. Phys. A*, 33 2997–3019.

Jammer, M., 1974. *The Philosophy of Quantum Mechanics*. Wiley, New York.

Jones, V. F. R., 1987. "Hecke algebra representations of braid groups and link polynomials," *Annals of Mathematics*, 126(2), 335–88.

Jordan, P., and E. Wigner, 1928. "Über das Paulische Äquivalenzverbot," *Zeitschrift für Physik*, 47(9), 631–51.

Kachru, S., R. Kallosh, A. Linde, and S. P. Trivedi, 2003. "de Sitter vacua in string theory," *Phys. Rev.* D, 68, 1841–56.

Kaluza, T., 1921. "Zum Unitätsproblem in der Physik," *Sitzungsber. Preuss. Akad. Wiss.* Berlin, (Math. Phys.), 96, 966–72.

Kauffman, L. H., 2001. *Knots and Physics*. World Scientific, Singapore.

Khlebnikov, S. Yu., and M. E. Shaposhnikov, 1996. "Melting the Higgs vacuum: Conserved numbers at high temperature," *Phys. Lett.* B, 387, 817–22.

Klein, F., 1926. *Vorlesungen über die Entwicklungen der Mathematik im 19. Jahrhundert*. Springer, Berlin.

Klein, O., 1926. "Quantentheorie und fünfdimensionale Relativitätstheorie," *Zeitschrift für Physik*, 37(12), 895–906.

Kobayashi, S., and K. Nomizu, 1969. *Foundations of Differential Geometry*. Interscience, New York.

Kovalenko, A.V., et al., 2007. "On topological properties of vacuum defects in lattice Yang-Mills theories," *Physics Letters B*, 648(5–6), 383–87.

Koyré, A., 1957. *From the Closed Word to the Open Universe*. Johns Hopkins Press, Baltimore.

Kreimer, D., 2000. *Knots and Feynman Diagrams*. Cambridge University Press, Cambridge.

Lai, C. H., ed., 1981. *Selected Papers on Gauge Theory of Weak and Electromagnetic Interactions*. World Scientific, Singapore.

Lamb, Jr., W. E., and R. C. Retherford, 1947. "Fine structure of the hydrogen atom by a microwave method," *Phys. Rev.* 72(3), 241–43.

Lamoreaux, S. K., 1997. "Demonstration of the Casimir force in the 0.6 to 6 μm range," *Phys. Rev. Lett.*, 78(1), 5–8.

Lee, T. D., 1990. *Particle Physics and Introduction to Field Theory*. Harwood Academic Publishers, London and New York.

Lee, T. D., and G. C. Wick, 1974. "Vacuum stability and vacuum excitation in a spin-0 field theory," *Phys. Rev.* D, 9, 2291–2316.

Leinaas, J. M., and K. Olaussen, 1986. "Vacuum structure and the fermion-boson transformation," *Phys. Rev.* D, 34(8), 2483–88.

Lemaître, G., 1933. "L'univers en expansion," *Ann. Soc. Sc. Bruxelles*, 53A, 51–83.

Levi-Civita, T., 1917. "Realtà fisica di alcuni spazi normali del Bianchi," *Rend. R. Acad. Lincei*, 26, 519–31.

Lévy-Leblond, J.-M., 2006. *De la Matière—Relativiste, Quantique, Interactive*. Éditions du Seuil, Paris.

Lima, J. A. S., and A. Maia, Jr., 1995. "On the thermodynamic properties of the quantum vacuum," *Intern. J. Theor. Phys.*, 34(9), 1835–42.

Linde, A., 1982. "A new inflationary universe scenario: A possible solution of the horizon, flatness, homogeneity, isotropy and primordial monopole problems," *Phys. Lett.* B, 108, 389–412.

Lorentz, H., 1909. *Theory of Electrons and Its Applications to the Phenomena of Light and Radiant Heat.* Columbia University Press, New York.

MacKenzie, R., and F. Wilczek, 1984. "Illustrations of vacuum polarization by solitons," *Phys. Rev.* D30, 2194–2217.

MacKenzie, R., F. Wilczek, and A. Zee, 1983. "Possible form of vacuum deformation by heavy particles," *Phys. Rev. Lett.*, 53, 2203–34.

Maldacena, J., 2007. "The illusion of gravity," *Sci. Amer. Rep.*, April, 75–81.

Manin, Y. I., 1988. *Gauge Field Theory and Complex Geometry.* Springer-Verlag, Berlin/ Heidelberg.

Maunder, G. R. F., 1980. *Algebraic Topology.* Cambridge University Press, Cambridge.

Maxwell, J. C., 1865. "A dynamical theory of the electromagnetic field," *Royal Society Transactions*, 155, p. 459.

Meissner, K. A., and G. Veneziano, 1991. "Symmetries of cosmological superstring vacua," *Phys. Lett.* B, 267, 33–48.

Mickelsoon, J., 1998. "Vacuum polarization and the geometric phase: Gauge invariance," *J. Math. Phys.*, 39(2), 831–37.

Milnor, J., 1966. *Topology from the Differentiable Viewpoint.* University of Virginia Press, Charlottesville.

Milonni, P. W., 1994. *The Quantum Vacuum: An Introduction to Quantum Electrodynamics.* Academic Press, San Diego.

Milonni, P. W., and M.-L. Shih, 1991. "Zero-point energy in early quantum theory," *Amer. J. Phys.*, 59(8), 684–98.

Minkowski, H., 1908/9. "Raum und Zeit," *Jahresberichte der Deutschen Mathematiker Vereinigung*, 18, 75–88.

Moffat, H., 1969. "The degree of unknottedness of tangled vortex lines," *J. Fluid Mech.*, 35, 117–29.

Mohideen, U., and A. Roy, 1998. "Precision measurement of the Casimir force," *Phys. Rev.*, 81, 4547–58.

Moschella, U., and R. Schaeffer, 1998. "Quantum fluctuations in the open universe," *Phys. Rev D*, 57, 2147–51.

———, 2007. "Quantum theory on Lobatchevski spaces," *Classical and Quantum Gravity*, 24(14), 3571–3602.

Nambu, Y., and G. Jona-Lasinio, 1961. "Dynamical model of elementary particles based on an analogy with superconductivity. I," *Phys. Rev.*, 122, 345–58.

Nash, C., and S. Sen, 1983. *Topology and Geometry for Physicists.* Academic Press, London.

Neumann, J. von, 1932. *Mathematische Grundlagen der Quantenmechanik.* Springer, Berlin.

Nordström, G., 1914. "Über die Möglichkeit, das eletromagnetische Feld und das Gravitationsfeld zu vereinigen," *Physikalische Zeitschrift*, 15, 504–6.

Novikov, I., and Ya. B. Zeldovitch, 1983. *The Structure and Evolution of the Universe.* University of Chicago Press, Chicago.

O'Raifeartaigh, L., 1997. *The Dawning of Gauge Theory.* Princeton University Press, Princeton.

O'Raifeartaigh, L., and N. Straumann, 1984. "On the origin of the Aharonov-Bohm effect," in M. Peshkin and A. Tonomura (eds.), *The Aharonov-Bohm Effect*. Springer-Verlag, Heidelberg.

Padmanabhan, T., 1983. "Quantum conformal fluctuations and stationary states," *Inter. J. Theor. Phys.*, 22(11), 1023–36.

Pais, A., 1986. *Inward Bound: Of Matter and Forces in the Physical World*. Oxford University Press, Oxford.

Parker, L., 1968. "Particle creation in exanding universes," *Phys. Rev. Lett.*, 21, 562–64.

———, 1969. "Quantized fields and particle creation in expanding universes," *Phys. Rev.*, 183, 1057–68.

Pascal, B., 1651. *Préface du traité du vide*, in *Œuvres complètes*, t. 2, GM Flammarion, Paris, 1985.

Passante, R., F. Persico, and L. Rizzuto, 2003. "Spatial correlations of vacuum fluctuations and the Casimir-polder potential," *Phys. Lett. A*, 316, 29.

Penrose, R., 1996. "On gravity's role in quantum state reduction," *Gen. Rel. Grav.*, 28, 581–600.

———, 2004. *The Road to Reality: A Complete Guide to the Laws of the Universe*. Vintage, London.

Penrose, R., and C. J. Isham (eds.), 1986. *Quantum Concepts in Space and Time*. Clarendon Press, Oxford.

Peshkin, M., and A. Tonomura, 1989. *The Aharonov-Bohm Effect*. Springer-Verlag, Berlin.

Peshkin, M. E., and D. V. Schroeder, 1995. *An Introduction to Quantum Field Theory*. Perseus Books, Cambridge, Mass.

Phillips, N. G., and B. L. Hu, 1997. "Fluctuations of the vacuum energy density of quantum fields in curved spacetime via generalized zeta functions," *Phys. Rev. D*, 55, 6123–34.

Planck, M., 1912. "Über die Begrundung des Gesetzes der Scharzen Strahlung," *Annalen der Physik*, 37, 642–59.

Podolny, R., 1986. *Something Called Nothing: Physical Vacuum: What Is It?* Mir Publishers, Moscow.

Politzer, H. D., 1970. "Reliable perturbative results for strong interactions," *Phys. Rev. Lett.*, 30(26), 1346–49.

———. 1974. "Asymptotic freedom: An approach to strong interactions," *Physics Reports*, 14, 129–80.

Polyakov, A. M., 1975. "Compact gauge fields and the infrared catastrophe," *Phys. Lett. B*, 59(1), 82–84.

Puthoff, H. E., 1989a. "Gravity as a zero-point-fluctuation force," *Phys. Rev. A*, 39(5), 2333–42.

———, 1989b. "Source of vacuum electromagnetic zero-point energy," *Phys. Rev. A*, 40(9), 4857–62.

Rafelski, J., and B. Müller, 1985. *Die Struktur des Vakuums*. Verlag H. Deutsch, Frankfurt.

Ramond, P., 1981. *Field Theory: A Modern Primer*. Benjamin Cummings, Reading, Mass.

Ray, S., 2006. "Some properties of meta-stable supersymmetry-breaking vacua in Wess-Zumino models," *Phys. Lett.* B, 642(1–2), 137–41.

Redhead, M. L. G., 1994. "The vacuum in relativistic quantum field theory," *Proc. Biennial Meeting, Philosophy of Science Association: PSA 1994,* M. F. D. Hull, M. Forbes, and R. M. Burian (eds.), vol. 2, 88–89.

————, 2003. "The interpretation of gauge symmetry," in M. Kuhlmann, H. Lyre, and A. Wayne (eds.), *Ontological Aspects of Quantum Field Theory.* World Scientific, Singapore.

Requardt, M., and S. Roy, 2001. "Quantum spacetime as a statistical geometry of fuzzy lumps and the connection with random metric spaces," *Classical and Quantum Gravity,* 18, 3039–53.

Reynaud, S., E. Giacobino, and J. Zinn-Justin (eds.), 1997. *Quantum Fluctuations,* Les Houches LXIII. Elsevier, London.

Reynolds, W., 1993. "Hyperbolic geometry on a hyperboloid," *Amer. Math. Monthly,* 100, 442–55.

Riemann, B., 1867. "Ueber die Hypothesen, welche der Geometrie zu Grunde liegen," *Abh. K. Gesell. Wiss. Gött.,* 13, 133–52 (*Habilitationsarbeit,* 1854).

Rindler, W., 1969. *Essential Relativity.* Van Nostrand Reinhold, New York.

Robertson, H. P., 1935. "Kinematics and world structure," *Astrophysical Journal,* 82, 284–301.

Rothwarf, F., and S. Roy, 2007. "Quantum vacuum and matter–antimatter cosmology," arXiv/astro-ph/0703280v3 12 Mar.

Rovelli, C., and L. Smolin, 1988. "Knot theory and quantum gravity," *Phys. Rev. Letters,* 61(10), 1155–58.

Roy, S., 1998. *Statistical Geometry and Applications to Microphysics and Cosmology.* Kluwer Academic Publishers, Dordrecht.

Rudin, W., 1987. *Real and Complex Analysis.* McGraw-Hill, New York.

Rueda, A., and B. Haisch, 2005. "Gravity and the quantum vacuum inertia hypothesis," *Ann. Phys.,* 14(8), 479–98.

Rugh, S. E., and H. Zinkernagel, 2002. "The quantum vacuum and the cosmological constant problem," *Studies in History and Philosophy of Modern Physics,* 33, 663–705.

Russell, B., 1951. *A Critical Exposition of the Philosophy of Leibniz,* with a new introduction by J. G. Slater. Routledge, London.

Sakharov, A. D., 1968. "Vacuum quantum fluctuations in curved space and the theory of gravitation," *Dokl. Akad. Nauk. SSSR,* 12, 1040–57.

Sakurai, J. J., 1985. *Modern Quantum Mechanics.* Addison-Wesley, Redwood City, Calif.

Salam, A., 1982. "On Kaluza-Klein theory," *Annals of Physics,* 141, 316–52.

————, 1990. *Unification of Fundamental Forces.* Cambridge University Press, Cambridge.

Sarangi, S., and S.-H. H. Tye, 2003. "A note on the quantum creation of universes," hep-th/0603237.

Saunders, S., and H. R. Brown (eds.), 2002. *The Philosophy of Vacuum.* Oxford University Press, Oxford.

Scharf, G., 1996. "Vacuum stability in quantum field theory," *Il Nuovo Cimento A*, 109(11), 1605–07.

Schmidt, M. G., 1987. "Instantons and the vacuum condensates of the SUSY-gauge theories," *Acta Physica Polonica*, B18 (9), 775–92.

Schrödinger, E., 1939. "The proper vibrations of the expanding universe," *Physica*, 6, 899–912.

———, 1956. *Expanding Universes*. Cambridge University Press, Cambridge.

Schwinger, J., 1948. "Quantum electrodynamics. I. A covariant formulation," *Phys. Rev.*, 74, 1439–61.

———, 1951. "On gauge invariance and vacuum polarization," *Phys. Rev.*, 82, 664–79.

———, 1958. *Selected Papers on Quantum Electrodynamics*. Dover, New York.

Shuryak, E. V., 2004. *The QCD Vacuum, Hadrons and Superdense Matter*. World Scientific Publishing, Singapore.

Sihna, K. P., C. Sivaram, and E. C. G. Sudarshan, 1976. "The superfluid vacuum state, time-varying cosmological constant, and non-singular cosmological models," *Foundations of Physics*, 6, 717–26.

Sitenko, Yu. A., 1997. "Nonlocality, self-adjointness, and Θ-vacuum in quantum field theory in spaces with nontrivial topology," arXiv:hep-th/9702148v1 20 Feb.

———, 2000. *Vacuum Polarization Effects in the Background of Nontrivial Topology*. Lecture Notes in Physics, vol. 543. Springer-Verlag, Berlin/Heidelberg.

Sitenko, Yu. A., and D. G. Rakityansky, 1998. "Vacuum energy induced by an external magnetic field in a curved space," *Phys. Atom. Nucl.*, 61, 790–801.

Skyrme, T. H. R., 1961. "A non-linear field theory," *Proc. Roy. Soc. A* (London), 260, 127–38.

Smith, Q., 1988. "The uncaused beginning of the universe," *Philosophy of Science*, 55(1), 39–57.

———, 1997. "The ontological interpretation of the wave function of the universe," *The Monist*, 80(1), 160–85.

Solomon, D., 1998. "Gauge invariance and the vacuum state," *Can. J. Phys.*, 76(2), 111–27.

Sorensen, R., 2006. "Nothingness," in *Stanford Encyclopedia of Philosophy*, 1–22. Stanford University Press, Stanford.

Sparnaay, M., 1958. "Measurements of attractive forces between flat plates," *Physica*, 24 751–64.

Starobinsky, A. A., and Ya. B. Zel'Dovich, 1988. "The spontaneous creation of the Universe," *Sov. Sci. Rev. Sect. E*, 6(2), 103–44.

Stokes, G. G., 1880. *Mathematical and Physical Papers*, vol. 1. Cambridge University Press, Cambridge.

Straumann, N., 2003. "On the cosmological constant problems and the astronomical evidence for a homogeneous energy density with negative pressure," pp. 7–52 in *Poincaré Seminar 2002: Vacuum Energy—Renormalization*, B. Duplantier and V. Rivasseau (eds.). Birkhäuser, Basel.

Streeruwitz, E., 1975. "Vacuum fluctuations of a quantized scalar field in a Robertson-Walker universe," *Phys. Rev.* D11(12), 3378–83.

————, 1977. "Effective Lagrangian for general relativity in presence of a quantized scalar field," *Phys. Lett. B*, 67(2), 210–19.

Strocchi, F., and A. Wightman, 1974. "Proof of the charge superselection rule in local relativistic quantum field theory," *J. Math. Phys.*, 15, 2198–2224.

Taubes, C. H., 1990. "Casson's invariant and gauge theory," *Journal of Differential Geometry*, 31, 547–99.

Thom, R., 1954. "Quelques propriétés globales des variétés différentiables," *Comm. Math. Helv.*, 28, 17–86.

't Hooft, G., 1976. "Computation of the quantum effect due to a four-dimensional pseudoparticle," *Phys. Rev. D*, 14, 3432–50.

't Hooft, G., 1981. "Topology of the gauge condition and new confinement phases in non-Abelian gauge theories," *Nucl. Phys. B*, 190, 455–78.

————, 1997. *In Search of the Ultimate Building Blocks*. Cambridge University Press, Cambridge.

————, 2006. "The conceptual basis of quantum field theory," in *Handbook of the Philosophy of Science, Philosophy of Physics*, D. M. Gabbay, P. Thagard, and J. Woods (eds.). Elsevier, London.

————, 2009. "The fundamental nature of space and time," pp. 13–25 in *Approaches to Quantum Gravity, Toward a New Understanding of Space, Time and Matter*, D. Oriti (ed.). Cambridge University Press, Cambridge.

Tomonaga, S., 1946. "On relativistically invariant formulation of the quantum theory of wave fields," *Progress of Theoretical Physics*, 1, 27–42.

————, 1948. "On infinite field reactions in quantum field theory," *Phys. Rev.*, 74, 224–25.

Tonomura, A., et al., 1986. "Evidence for Aharonov-Bohm effect with magnetic field completely shielded from electron wave," *Phys. Rev. Lett.*, 56, 792–95.

Trautman, A., and I. Robinson, 1960. "Spherical gravitational waves," *Phys. Rev. Lett.*, 4, 431–32.

Tryon, E. P., 1973. "Is the universe a vacuum fluctuation?," *Nature*, 246, 396–97.

Turner, M. S., 2003. "Dark energy and the destiny of the universe," pp. 127–37 in *Poincaré Seminar 2002: Vacuum Energy, Renormalization*, B. Duplantier and V. Rivasseau (eds.). Birkhäuser, Basel.

Turner, M. S., and F. Wilczek, 1982. "Is our vacuum metastable," *Nature*, 298, 633–37.

Utiyama, R., and B. S. DeWitt, 1962. "Renormalization of a classical gravitational field interacting with quantized matter," *J. Math. Phys.*, 3, 608–19.

Van Baal, P., 2002. "Reflections on topology—viewpoints on Abelian projections," *Nuclear Physics B* (Proc. Suppl.), 108, 3–11.

Van Baal, P., and A. Wipf, 2001. "Classical gauge vacua as knots," *Phys. Lett. B*, 515, 181–84.

Veltman, M. J. G., 2003. *Facts and Mysteries in Elementary Particle Physics*. World Scientific, London.

Veneziano, G., 1968. "Construction of crossing-symmetric, Regge behaved amplitude for linearly rising trajectories," *Nuovo Cimento*, 57A, 190–97.

————, 1989. "Wormholes, non-local actions and a new mechanism for suppressing the cosmology constant," *Phys. Lett. B*, 228, 210–31.

————, 1998. "Quantum geometric origins of all forces in string theory," *The Geometric Universe: Science, Geometry, and the Work of Roger Penrose*, S. A. Huggett et al., (eds.), (235–43). Oxford University Press, Oxford.

Vilenkin, A., 1982. "Creation of universes from nothing," *Phys. Lett. B*, 117(1–2), 25–28.

————, 1984. "Quantum creation of universes," *Phys. Rev. D*, 30(2), 509–11.

————, 1998. "The quantum cosmology debate," gr-qc/9812027v1.

Walker, A. G., 1937. "On Milne's theory of world-structure," *Proc. London Mathematical Society*, 42(2), 90–127.

Warner, F., 1983. *Foundations of Differential Manifolds and Lie Groups*. Springer-Verlag, Berlin.

Wasson, P. S., 2006. *Five-Dimensional Physics: Classical and Quantum Consequences of Kaluza-Klein Cosmology*. World Scientific, Singapore.

Weinberg, S., 1978. *Gravitation and Cosmology: Principles and Applications of General Relativity*. Wiley, New York.

————, 1989. "The cosmological constant," *Rev. Mod. Phys.*, 61, 1–23.

————, 1995. *The Quantum Theory of Fields*. Cambridge University Press, Cambridge.

Weisskopf, V., 1934. "Über die Selbstenergie des Elektrons," *Zeit. Phys.*, 89, 27–39.

————, 1939. "On the self-energy and the electromagnetic field of the electron," *Phys. Rev.*, 56, 72–85.

Wess, J., and B. Zunino, 1974. "A Lagrangian model invariant under supergauge transformations," *Phys. Lett. B*, 49, 52–75.

Weyl, H., 1918. *Raum-Zeit-Materie*, Springer-Verlag, Berlin.

————, 1928. *Gruppentheorie und Quantenmechanics*. Springer-Verlag, Leipzig.

Wheeler, J. A., 1962. *Geometrodynamics*. Academic Press, New York.

————, 1968. "Superspace and the nature of quantum geometrodynamics," in *Battelle Rencontres*, 1967 Lectures in Mathematics and Physics, C. M. DeWitt and J. A. Wheeler (eds.), 243–307. Benjamin, New York.

Wigner, E. P., 1939. "The unitary representation of the inhomogeneous Lorentz group," *Annals of Mathematics*, 40, 149–87.

————, 1967. *Symmetries and Reflections*. Indiana University Press, Bloomington and London.

Wilczek, F., 1999. "Quantum field theory," *Reviews of Modern Physics*, 71(2), 83–95.

————, 2000. "QCD made simple," *Physics Today*, August, 22–27.

————, 2003. "QCD and natural philosophy," *Annales Institut Henri Poincaré*, 4, 211–28.

Wilczek, F., and D. J. Gross, 1973. "Asymptotically free gauge theories. I," *Phys. Rev. D*, 8, 3633–52.

Wilczek, F., and Y.-S. Wu, 1990. "Space-time approach to holonomy scattering," *Phys. Rev. Lett.*, 65, 13–32.

Wilczek, F., and A. Zee, 1983. "Linking numbers, spin, and statistics of solitons," *Phys. Rev. Lett.*, 51, 2250–64.

Witten, E., 1981. "Cosmological consequences of a light Higgs boson," *Nucl. Phys. B*, 177, 477–88.

————, 1987. "Physics and geometry," pp. 267–303 in Proc. International Congress of Mathematicians, Berkeley, American Mathematical Society.

——, 1988. "Topological quantum field theory," *Comm. Math. Phys.*, 117, 353–86.

——, 1989. "Quantum field theory and the Jones polynomial," *Comm. Math. Phys.*, 121, 351–99.

——, 2001. "Black holes and quark confinement," *Current Science*, 81 (12), 1576–81.

Wu, T. T., and C. N. Yang, 1975. "Concept of non-integrable phase factors and global formulation of gauge fields," *Phys. Rev. D*, 12, 3845–57.

Wuensch, D., 2003. "The 5th dimension: Theodor Kaluza's groundbreaking idea," *Annalen der Physik*, 12(9), 519–42.

Yang, C. N., 1980. "Hermann Weyl's contribution to physics," in *Hermann Weyl*, K. Chandrasekharan (ed.). Springer-Verlag, Berlin.

——, 1983. "Symmetry principles in modern physics," in *Selected Papers, C. N. Yang, 1945–1980*. W. F. Freeman, San Francisco.

Yang, C. N., and R. Mills, 1954. "Conservation of isotopic spin and isotopic spin invariance," *Phys. Rev.* 96(1), 191–95.

Young, N., 1998. *An Introduction to Hilbert Space*. Cambridge University Press, Cambridge.

Zel'dovich, Ya. B., 1976. "Charge asymmetry of the universe as a consequence of evaporation of black holes and the asymmetry of the weak interaction," *JETP Letters*, 24, 25–28.

——, 1979. "Cosmology and the early universe," in *General Relativity. An Einstein Centenary Survey*, pp. 518–32 in S. W. Hawking and W. Israel (eds.). Cambridge University Press, Cambridge.

Zel'dovich, Ya. B., and I. D. Novikov, 1967. "Relativistic Astrophysics," *Sov. Astron. AJ.*, 11, 526–30.

Zumino, B., 1975. "Supersymmetry and the vacuum," *Nuclear Physics B*, 89(3), 535–46.

Zwanziger, D., 1976. "Physical states in quantum electrodynamics," *Phys. Rev. D*, 14, 2570–89.

Index